Motion Structures

Motion structures are simply assemblies of resistant bodies connected by movable joints. Unlike conventional structures, they allow large shape transformations to satisfy practical requirements and they can be used in

- shelters, emergency structures and exhibition stands,
- aircraft morphing wings,
- satellite solar panels and space antennas,
- morphing core materials for composites,
- medical implants for minimum invasive surgery.

Though traditionally the subject falls within structural engineering, motion structures are more closely related to mechanisms, and they draw on the principles of kinematic and geometrical analysis in their design. Indeed their design and analysis can be viewed as an extension of the theory of mechanisms, and can make effective use of a wealth of mathematical principles.

This book outlines the relevant underlying theory and motion structural concepts, and uses a number of innovative but simple structures as examples.

Zhong You is a Fellow of Magdalen College, Oxford and lecturer at Oxford University, UK.

Yan Chen is an Assistant Professor at Nanyang Technological University, Singapore.

Motion Structures

Deployable structural assemblies of mechanisms

Zhong You and Yan Chen

Spon Press
an imprint of Taylor & Francis
LONDON AND NEW YORK

First published 2012
by Spon Press
2 Park Square, Milton Park, Abingdon, Oxon OX14 4RN

Simultaneously published in the USA and Canada
by Spon Press
711 Third Avenue, New York, NY 10017

First issued in paperback 2019

Routledge is an imprint of the Taylor & Francis Group, an informa business

British Library Cataloguing in Publication Data
A catalogue record for this book is available from the British Library

Library of Congress Cataloging in Publication Data
You, Zhong, 1963–
Motion structures : deployable structural assemblies of mechanisms /
Zhong You and Yan Chen.
  p. cm.
  Includes bibliographical references.
  1. Flexible structures. 2. Buildings–Joints. 3. Buildings–Mechanical
  equipment. 4. Mechanical movements. I. Chen, Yan, 1974- II. Title.
  TA660.F53Y68 2011
  624.1'7–dc22

                                                            2011003147

ISBN-13: 978-0-415-55489-3 (hbk)
ISBN-13: 978-0-367-86557-3 (pbk)

DOI: 10.1201/9781482266610

Typeset in Sabon
by Wearset Ltd, Boldon, Tyne and Wear

Contents

Acknowledgements

The authors are indebted to many people who greatly influenced the research that eventually led to this book. First and foremost we would like to thank Professor Sergio Pellegrino for guiding us into this exciting inter-disciplinary area of deployable structures two decades ago. Many motion structures introduced herein were originated from his pioneering work. The authors are also grateful to Professors Christopher Calladine and Tibor Tarnai, whose insight into many of the problems that we have worked on and encouragement have inspired us to seek more elegant analytical solutions. Our collaborators – Nicholas Cole, Paul Kassabian, Alan Kwan, Shiyu Liu, Yaozhi Luo, Decan Mao and Chaoyang Song, to name just a few – have undoubtedly had an impact on the approach that has been taken. The joint work has been extensively cited herein. Our gratitude also goes to Professor Guy Houlsby for his continuous support, to the workshop of the Department of Engineering Science, University of Oxford, for assistance in making the wonderful models, to Jennet Hovard and John Mooney who produced the most of the superb photos for this book, and to Joseph Gattas and Alison May who patiently read the first draft of the book. Finally we would like to acknowledge the help, patience and encouragement provided by our editors at Spon Press, Tony Moore, Siobhán Poole and Simon Bates.

1 Introduction

DOI: 10.1201/9781482266610-1

A structure is a combination of resistant bodies made to bear loads. In general no internal mobility or relative motions among its members are allowed. However, there exists a family of unconventional structures, found in many places from common household items such as umbrellas and foldable chairs to solar panels of spacecrafts and retractable roofs, that are capable of large shape changes. These structures are commonly known as deployable structures. The purpose of adopting deployable structures is to have convenience in transportation or storage, but some, e.g. the retractable roofs, are concerned primarily with provision of instant coverage or shelter to create a desirable environment. In accordance with the deployment process, they fall into two categories. The first category is the deformable structure characterised by the fact that the overall strain energy of the structure varies during geometrical transformation. Typical examples include the inflatable structures such as balloons and cardiovascular stents, a type of medical device placed via minimum invasive surgery for treatment of blockage in blood vessels. The other category is essentially mechanism. The deployment is executed by activation of one or a number of carefully designed internal mechanisms. Retractable roofs for sports facilities and a toy called the Hoberman sphere belong to the second category.

This book focuses on the second category. The term *Motion Structures* is adopted to represent this branch of the deployable structure family owing to the existence of internal mechanisms.

A mechanism in machine theory, which is referred to as a conventional mechanism hereafter, is commonly identified as a set of moving or working parts used essentially as a means of transmitting motions or controlling movement of one part relative to another. It is often assembled from gears, cams and linkages, though it may also contain other specialised components, e.g. springs, ratchets, brakes, and clutches, etc. There are close similarities as well as distinct differences between a motion structure and a conventional mechanism. First, the primary function of a motion structure is to have shape alteration essential to practical requirements, rather than transmitting or controlling motions. Second, as a structure, a motion structure is usually composed of far more parts than a conventional mechanism.

Third, the motion structures generally use fewer but more robust types of joints because of the environments in which they typically operate. Moreover, when synthesising the motion structures, the positions and orientations of the parts during the motion are far more important than other physical properties such as velocity and acceleration, as the cycle time of a motion structure is generally in a matter of minutes rather than seconds or less for conventional mechanisms.

There are two ways to synthesise motion structures. Structural engineers and architects, responsible for most of the motion structures for civil engineering applications such as retractable roofs and façades, are inclined to adopt concepts involving a small number of larger moving bodies whose motions are then synchronised by electronic means (Ishii, 2000). Each moving body may consist of many members but the internal relative motions among the members are prohibited. The reason for taking this approach is because of the difficulty in assembling a large number of members together while retaining the internal mobility. This approach becomes less effective when the purpose of motion structures is to achieve small and compact packaging size. An alternative is to select known conventional mechanisms as basic building blocks and then to assemble such blocks together in such a way that the degrees of freedom of each mechanism are retained. For example, the design of an umbrella frame can be regarded as a combination of a number of identical linkages, each of which is a supporting frame with three hinges and one slider. The sliders are then merged at the central pole so that the entire frame has one degree of freedom.

In the past few years, the authors have created a number of novel motion structures by the latter approach. Here we are to share with readers our experience. Almost all of the motion structures in the book are one degree-of-freedom assembly created by basic mechanisms such as the planar and spatial linkages with only hinge joints.

In the following pages, the reader will first encounter in Chapter 2 the terms and definitions in commonly used mechanism theory, mobility criteria and analytical methods for synthesising mechanisms, some basic mechanisms and a link between mechanisms and structures. The basic mechanisms are then used to construct planar motion structures in Chapter 3, followed by rings and domes in Chapter 4. Chapters 5 and 6 are about truly three dimensional motion structures using spatial linkages as building blocks. A method that uses tilings for designing the layouts of three dimensional motion structures is given in Chapter 7, which ends the book.

The book is limited to the geometrical aspects of motion structures. In analysis and design synthesis, it is always assumed that the motion structures are made of *rigid* bodies. It should be clear that the rigid body is only an approximate reality in practice. This assumption reduces the motion of a motion structure to geometry. This is true if the bearing clearance and elastic deformation of members are sufficiently small. The motions of a geometry concept are reproduced in the actual structure without noticeable error.

The scope of the book means that it alone is not complete in a practical engineering sense. To arrive at the final design of a motion structure, one must decide plenty of details not covered by this book. For instance, one must choose materials, carry out structural analysis for a series of configurations subjected to the expected loadings, consider influences of dynamic loadings such as inertia, and decide other dimensions of the components, how the members are manufactured and assembled and how the motion structures are to be serviced including lubrication of the bearings and design of their clearances to ensure that the mechanism will remain throughout its working life strong enough and stiff enough to withstand all the forces it will experience.

The design of motion structures remains an iterative process despite the best efforts of the authors and many fellow engineers. However, it remains to be our strong belief that if the scope of motion structures is to be extended, engineers who create such structures must become familiar with theoretical methods of synthesis, with criteria which must be obeyed and with general principles of motion which can always provide shortcuts across the tedious processes of identifying a starting concept. A sound knowledge in mechanisms, combined with common sense and intuition, can contribute to the production of many more efficient motion structures.

2 Fundamental concepts, methods and classification

DOI: 10.1201/9781482266610-2

2.1 Introduction

2.1.1 Definitions

In this book a *mechanism* is defined as the assembly of rigid members, also known as *links*, connected by kinematic joints. A mechanism is sometimes referred to as a *kinematic chain* as well.

A *kinematic joint* is formed by direct contact between the surfaces of two members. Joints are the most important aspect of a mechanism in the view of kinematics. They permit relative motion in some directions while constraining motion in others. The types of motion permitted are related to the *number of degrees of freedom* of the joint which is equal to the minimum number of independent coordinates needed to uniquely specify the position of a link relative to the other constrained by the joint. Reuleaux (1875) published the first book on theoretical kinematics of mechanisms, in which he called a kinematic joint a *pair*. He further divided joints into lower pairs and higher pairs. A *lower pair* is the one in which contact between two rigid members occurs at every point of one or more surface segments. A *higher pair* is one in which contact occurs only at isolated points or along line segments.

Due to the requirement of surface contact, there are only six fundamentally different types of lower pairs classified by the types of relative motions that they permit, all of which are listed in Table 2.1. There are, in contrast, an infinite number of possible higher pair geometries. A couple of examples of the higher pairs are given in Table 2.2.

A kinematic chain is commonly known as a *linkage* if it is made from a series of links connected by only lower pair joints. Whenever there is one higher pair or more, it must be called a mechanism and should not be placed in the linkage sub-class.

2.1.2 Mobility

The number of degrees of freedom of a mechanism is normally called the *mobility*, which is the number of inputs required to determine the position

Table 2.1 The lower pair joints

Joint name	Letter symbol	Number of degrees of freedom	Typical form	Sketch symbol
Revolute joint (hinge, turning pair or pin)	R	1		
Prismatic joint (slider or sliding pair)	P	1		
Screw joint (helical joint or helical pair)	H	1	$s=k\theta$	
Cylindrical joint (cylindrical pair)	C	2		
Spherical joint (ball joint or spherical pair)	S	3		
Planar joint (planar pair)	P_L	3		

of all the links, also known as outputs, with respect to a fixed reference frame, sometimes referred to as *ground*. The mechanism becomes a locked chain, or a conventional structure, if no mobility remains.

In three dimensional space, the number of degrees of freedom of a rigid link is six: three directional displacements and three directional rotations. Thus n free links will have $6n$ degrees of freedom. By fixing one link as ground, the remaining degrees of freedom are $6(n-1)$. A joint with f degrees of freedom connecting two links reduces the total degrees of

Table 2.2 Some of the higher pair joints

Joint names	Number of degrees of freedom	Typical form	Comments
Cylindrical roller	1	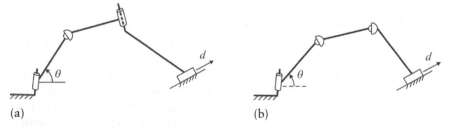	Roller rotates about the instantaneous contact line and does not slip on the surface.
Spatial point contact	5		Body can rotate about any axis through the contact point and slide in any direction in the tangent plane.

freedom by $6-f$. For a mechanism composed of n links that are connected with a total of j joints, each of which has f_i ($i=1, 2,\ldots, j$) degrees of freedom, the mobility of the linkage, m, is

$$6(n-1)-\sum_{i=1}^{j}(6-f_i),$$

or

$$m = 6(n-j-1)+\sum_{i=1}^{j}f_i \tag{2.1}$$

This is called the *Grübler–Kutzbach mobility criterion* (Hunt, 1978) or *Kutzbach criterion*.

Figure 2.1(a) shows a spatial linkage composed of a chain of four links and four joints, including one revolute joint (R), one spherical joint (S), one cylindrical joint (C) and one prismatic joint (P), which is commonly referred to as an *RSCP* linkage. According to Eq. (2.1), its mobility is

$$m = 6(n-j-1)+\sum_{i=1}^{j}f_i = 6(4-4-1)+(1+3+2+1)=1.$$

(a)   (b)

Figure 2.1 (a) an *RSCP* linkage and (b) an *RSSP* linkage.

The output displacement at the prismatic joint is therefore uniquely decided by the input angle at the revolute joint.

Due to the fact that the Kutzbach criterion given in Eq. (2.1) considers only the topological information of a linkage, it can sometimes give misleading results. Take the *RSSP* linkage in Figure 2.1(b) as an example. It has four links and four joints: one revolute, two spherical and one prismatic joint. Therefore,

$$m = 6(n - j - 1) + \sum_{i=1}^{j} f_i = 6(4 - 4 - 1) + (1 + 3 + 3 + 1) = 2 \cdot$$

This contradicts with the fact that there is a unique output displacement at the prismatic joint for any given value of the input variable at the revolute joint. Then, where is the second mobility? Examination of the linkage reveals that the coupler link that bridges two spherical joints is free to spin about the line through the centres of the spherical joints. Such motion, called *an idle degree of freedom*, can take place in any position of the linkage without affecting the real input and output relationship. So when idle degrees of freedom exist the Kutzbach criterion gives the result that the number of mobility is greater than the number of useful degrees of freedom.

Figure 2.2 shows a tetrahedron truss. It has a total of six rigid bars (links) and four spherical joints, each of which connects three bars. So $n = 6$ and $j = 8$ because in the Kutzbach criterion a connection between a pair of links is counted as one joint and hence there are two joints when three links are connected to a spherical joint. Each spherical joint has three degrees of freedom. So Eq. (2.1) gives

$$m = 6(6 - 8 - 1) + (3 \times 8) = 6 \cdot$$

Here, the six degrees of freedom refer to the rotation of each bar about its respective axis linking two joints, all of which are idle degrees of freedom. If one of six bars is taken as the ground, the entire tetrahedron truss is able to rotate about it.

In addition to *idle degrees of freedom*, the Kutzbach criterion may also yield result that the mobility is smaller than the number of actual degrees of freedom. This situation will be examined in detail in Section 2.3.

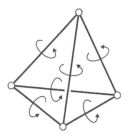

Figure 2.2 A tetrahedron truss.

In general, mechanisms are three dimensional. However, there are circumstances where the mechanisms, and in particular the way they are analysed, can be simplified. This leads to two specific types of mechanisms, namely the planar and spherical mechanisms.

2.1.3 The planar mechanisms

A *planar mechanism* is a mechanism such that the trajectories of all points on its links are parallel to a plane. The plane is known as the *plane of motion*. The only lower pair joints that are properly compatible with planar motion are revolute and prismatic joints. The axes of rotation of all revolute joints must be normal to the plane of motion and the directions of sliding of all prismatic joints must be parallel to the plane of motion. And only cam and gear pairs as higher pair joints can be applied to planar mechanisms.

In a plane, each free rigid body has three degrees of freedom. The general mobility criterion (2.1) therefore becomes

$$m = 3(n - j - 1) + \sum_{i=1}^{j} f_i. \tag{2.2}$$

A planar mechanism becomes a *planar linkage* if it contains only revolute and prismatic joints, which is also the focus of this book for they can be effectively utilised to construct large motion assemblies. Because both revolute and prismatic joints have one degree of freedom, i.e. $f_i = 1$, the mobility criterion of planar linkage becomes

$$m = 3(n-1) - 2j \tag{2.3}$$

Applying Eq. (2.3) to a simple bathroom retractable mirror, Figure 2.3, which has twelve links (eight bars, two vertical members and two sleeves), fourteen revolute joints and two prismatic joints, yields

$$m = 3(n-1) - 2j = 3 \times (12-1) + 2 \times 16 = 1,$$

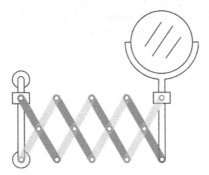

Figure 2.3 A bathroom retractable mirror.

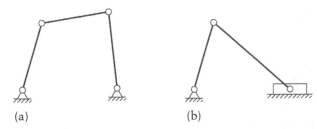

Figure 2.4 (a) A planar 4R linkage and (b) a slider-crank linkage.

which corresponds to the retraction and expansion of the mirror frame.

Some planar linkages have rigid links joined together to form a single closed chain. In this situation, $n=j$, and thus at least four links ($n=4$) are needed to achieve mobility one ($m=1$) according to Eq. (2.3). When the four links are connected by four revolute joints (R), Figure 2.4(a), we obtain a planar 4R linkage, also called a four-bar linkage. Connecting four links with three revolute joints and one prismatic joint (P) results in a planar *RRRP* linkage, Figure 2.4(b), which is also called a slider-crank linkage. In both linkages, ground is counted as one link.

2.1.4 The spherical mechanisms

The spherical mechanism is a mechanism where all of the links are constrained to rotate about the same fixed point in space. The trajectories of points on the links therefore lie on concentric spheres. In general spherical mechanisms include not only the linkages with revolute joints and arc prismatic joints, but also spherical cam mechanisms and bevel gears, and tapered roller bearing.

The most useful spherical mechanism for the purpose of constructing large motion structures is the spherical linkage: a closed chain of links joined together by revolute joints whose axes meet at one point. The spherical linkage is much like the planar linkage for all of the revolute joint axes are parallel in a planar linkage, whereas in a spherical linkage they intersect at a point known as the concurrency point. In fact, a planar linkage can be deemed as a spherical linkage for which the concurrency point is at infinity.

There are many similarities in the properties of spherical and planar linkages. Due to the concurrency constraint, the number of degrees of freedom for a rigid body in the spherical space is three, which are rotations about three perpendicular axes passing the concurrency point. Therefore, the Kutzbach criterion has a particular form for the spherical linkage as it is for the planar linkage, which is

$$m = 3(n-1) - 2j. \qquad (2.4)$$

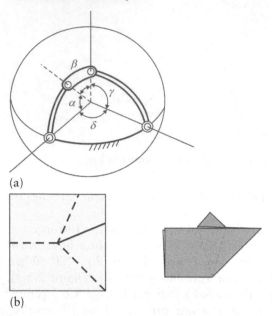

(a)

(b)

Figure 2.5 (a) A 4R spherical linkage and (b) an origami pattern with one
vertex and four creases.

Consequently, the simplest nontrivial spherical linkage is a four-bar spherical linkage, see Figure 2.5(a). The most convenient way to construct a
spherical linkage is by folding a piece of card with four creases, Figure
2.5(b): three valley creases, shown in dash lines (paper folds forward on to
itself) and one mountain crease, shown in solid line (paper folds away
from itself), that meet at a vertex. The creases act as revolute joints.

2.2 Kinematics of linkages

Kinematics studies the geometric properties of the motion of mechanisms,
including the positions, velocities and accelerations of points on the links
without consideration of the forces that cause the motion. A number of
kinematic methods have been developed in the past, including the matrix
method (Denavit and Hartenberg, 1955; Hartenberg and Denavit, 1964;
Beggs, 1966), quaternion and duel quaternion method (Altmann, 1986;
Kuipers, 2002; McCarthy, 1990), screw theory (Ball, 1876; McCarthy,
1990), Lie group and Lie algebra (Varadarajan, 1974), some of which can
be advantageous in analysing particular groups of mechanisms and in
finding specific physical quantities. For the purpose of design shape chang-
ing assemblies it is vital to identify the positions and angular positions of
the links in motion whereas the other physical quantities are of less

interest. Hence, here we shall concentrate on the kinematic analysis methods that provide the essential information.

2.2.1 *The vector method with complex numbers*

The simplest and most effective method for the analysis of planar mechanisms is the vector method involving complex numbers, which was first introduced by Bloch (Beggs, 1966). In a planar mechanism shown in Figure 2.6(a), let [x, y] be a fixed reference frame with its origin at one end of link 1, Figure 2.6(b). Link 1 can be represented by a vector $\mathbf{p}_1$,

$$\mathbf{p}_1 = \begin{Bmatrix} x \\ y \end{Bmatrix} = p_1 \begin{Bmatrix} \cos\theta_1 \\ \sin\theta_1 \end{Bmatrix},$$

where p_1 is the length of $\mathbf{p}_1$. This can also be written in complex polar form with real and imaginary parts corresponding to the Cartesian coordinates

$$\mathbf{p}_1 = p_1 e^{j\theta_1},$$

in which j is the standard imaginary unit. The subsequent links, e.g. links 2 and 3, can be written as

$$\mathbf{p}_2 = p_2 e^{j\theta_2} \text{ and } \mathbf{p}_3 = p_3 e^{j\theta_3}.$$

In a mechanism where one or more closed chains can be identified, around any closed chain with n links there must be

$$\sum_{i-1}^{n} \mathbf{p}_i = \sum_{i=1}^{n} p_i e^{j\theta_i} = 0. \tag{2.5}$$

Eq. (2.5) is an important vector equation called *the closure equation*.

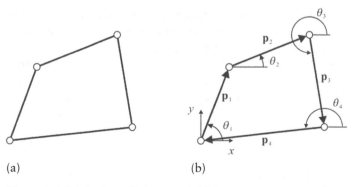

(a)   (b)

Figure 2.6 (a) A planar linkage and (b) its vector representation.

The closure equation is particularly useful in determining algebraically the relationship between the inputs and outputs. Take the four-bar linkage shown in Figure 2.6(b) as an example. Link 4 is chosen as ground and the links are represented by vectors p_i $(i = 1, 2, 3$ and $4)$. The closure equation (2.5) is

$$\mathbf{p}_1 + \mathbf{p}_2 + \mathbf{p}_3 + \mathbf{p}_4 = \mathbf{0}, \tag{2.6}$$

or

$$p_1 e^{j\theta_1} + p_2 e^{j\theta_2} + p_3 e^{j\theta_3} + p_4 e^{j\theta_4} = \mathbf{0}, \tag{2.7}$$

if the links are characterised by length p_i and angle with respect to x axis θ_i. The real and imaginary parts of Eq. (2.7) are

$$p_1 \cos\theta_1 + p_2 \cos\theta_2 + p_3 \cos\theta_3 + p_4 \cos\theta_4 = 0 \tag{2.8a}$$

$$p_1 \sin\theta_1 + p_2 \sin\theta_2 + p_3 \sin\theta_3 + p_4 \sin\theta_4 = 0 \tag{2.8b}$$

θ_4, the angle between the ground and fixed reference frame, is known. It is possible to determine outputs θ_2 and θ_3 by solving simultaneous equations (2.8a) and (2.8b) if θ_1 is taken as the input whose value is given.

It should be pointed out that the vector method using complex numbers is only suitable for modelling planar mechanisms. The three dimensional vector method can be applied to the spatial mechanism in statics and dynamics as well as the velocity and acceleration analysis when the position information is given. However, the complex notations will have to be replaced by the quaternions or dual quaternions. The matrix method to be introduced next is a much simpler alternative for kinematic analysis of spatial mechanisms.

2.2.2 The matrix method

The matrix method appears in most textbooks on mechanisms and is widely adopted for analysing spatial linkages as it requests the least amount of background knowledge in mathematics.

Let $[x_1, y_1, z_1]$ be a fixed reference frame and let $[x_2, y_2, z_2]$ be a reference frame fixed to the moving link, see Figure 2.7. The coordinate of the point P on the moving link in the fixed reference frame may be obtained from its coordinates in the moving reference frame by a transformation of the form

$$\mathbf{p}_1 = \mathbf{Q}\mathbf{p}_2 + \mathbf{q}_{12}, \tag{2.9}$$

where $\mathbf{p}_1 = [x_1, y_1, z_1]^T$, $\mathbf{p}_2 = [x_2, y_2, z_2]^T$, $\mathbf{Q}$ is 3×3 matrix related to the rotation angles of three axes of the moving reference frame relative to the fixed

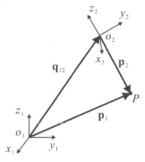

Figure 2.7 Vector transformation in space.

reference frame, and q_{12} is the vector related to the translation of the origin of the moving reference frame relative to the fixed reference frame.

We can apply this transformation model to a serial chain by establishing a convention that defines a consistent location for the reference frame in each member of the chain. The transformations representing the positions of the joints can then be applied successfully to produce the transformation relating to the end members of the chain.

Figure 2.8 shows the important geometric features of two adjacent links $(i-1)i$ and $i(i+1)$. Link $(i-1)i$ contains two revolute joints $(i-1)$ and i, whereas link $i(i+1)$ has two revolute joints i and $(i+1)$. They are connected by the revolute joint i. At each joint, a coordinate system is set up in such a way that z_i is the axis of revolute joint i, x_i is the axis commonly normal to z_{i-1} which is the axis of revolute joint $(i-1)$, and z_i, positively from joint $(i-1)$ to joint i, and y_i is the third axis following the right-hand rule. In the coordinate system, the geometric parameters of the links are defined as

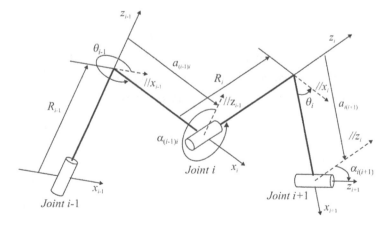

Figure 2.8 Coordinate systems for links connected by revolute joints.

follows. The *length* of link$(i-1)i$, $a_{(i-1)i}$, is the shortest distance between axes z_{i-1} and z_i; the *twist* of link $(i-1)i$, $a_{(i-1)i}$, is the angle of rotation from axes z_{i-1} to z_i positively about axis x_i; and the *offset* of joint i, R_i, is the distance from link $(i-1)i$ to link $i(i+1)$ positively about z_i. The kinematic parameter of the revolution of joints, the *revolute variable* of the linkage, θ_i, is the angle of rotation from x_{i-1} to x_i positively about z_i. Therefore, the rotation transfer matrix between the coordinate system of link $(i-1)i$ and that of link $i(i+1)$ is

$$\mathbf{Q}_{i(i+1)} = \begin{bmatrix} \cos\theta_i & \sin\theta_i & 0 \\ -\cos\alpha_{i(i+1)}\sin\theta_i & \cos\alpha_{i(i+1)}\cos\theta_i & \sin\alpha_{i(i+1)} \\ \sin\alpha_{i(i+1)}\sin\theta_i & -\sin\alpha_{i(i+1)}\sin\theta_i & \cos\alpha_{i(i+1)} \end{bmatrix}, \tag{2.10}$$

and the corresponding translation vector is

$$\mathbf{q}_{i(i+1)} = \begin{bmatrix} -a_{i(i+1)} \\ -R_i\sin\alpha_{i(i+1)} \\ -R_i\cos\alpha_{i(i+1)} \end{bmatrix}. \tag{2.11}$$

So we have, based on Eq. (2.9),

$$\mathbf{p}_{(i+1)} = \mathbf{Q}_{i(i+1)}\mathbf{p}_i + \mathbf{q}_{i(i+1)} \tag{2.12}$$

Denavit and Hartenberg (Beggs, 1966) introduced a notation to express Eq. (2.12) in a single four dimensional matrix-vector expression by using the homogeneous coordinates,

$$\mathbf{p}'_{(i+1)} = \mathbf{T}_{i(i+1)}\mathbf{p}'_i, \tag{2.13}$$

where

$$\mathbf{T}_{i(i+1)} = \begin{bmatrix} 1 & 0 & 0 & 0 \\ -a_{i(i+1)} & \cos\theta_i & \sin\theta_i & 0 \\ -R_i\sin\alpha_{i(i+1)} & -\cos\alpha_{i(i+1)}\sin\theta_i & \cos\alpha_{i(i+1)}\cos\theta_i & \sin\alpha_{i(i+1)} \\ -R_i\cos\alpha_{i(i+1)} & \sin\alpha_{i(i+1)}\sin\theta_i & -\sin\alpha_{i(i+1)}\cos\theta_i & \cos\alpha_{i(i+1)} \end{bmatrix} \tag{2.14}$$

and

$$\mathbf{p}'_i = \begin{bmatrix} 1 \\ x_i \\ y_i \\ z_i \end{bmatrix}. \tag{2.15}$$

It can be shown that the transfer matrix given in Eq. (2.14) also satisfies

$$\mathbf{T}_{(i+1)i} = \mathbf{T}_{i(i+1)}^{-1} = \begin{bmatrix} 1 & 0 & 0 & 0 \\ a_{i(i+1)}\cos\theta_i & \cos\theta_i & -\cos\alpha_{i(i+1)}\sin\theta_i & \sin\alpha_{i(i+1)}\sin\theta_i \\ a_{i(i+1)}\sin\theta_i & \sin\theta_i & \cos\alpha_{i(i+1)}\cos\theta_i & -\sin\alpha_{i(i+1)}\cos\theta_i \\ R_i & 0 & \sin\alpha_{i(i+1)} & \cos\alpha_{i(i+1)} \end{bmatrix}.$$

$$(2.16)$$

For a closed kinematic chain consisting of n links, by repeatedly applying Eq. (2.13), we have $\mathbf{p}'_2 = \mathbf{T}_{12}\mathbf{p}'_1$, $\mathbf{p}'_3 = \mathbf{T}_{23}\mathbf{p}'_2$, ..., $\mathbf{p}'_n = \mathbf{T}_{(n-1)n}\mathbf{p}'_{n-1}$ and $\mathbf{p}'_1 = \mathbf{T}_{n1}\mathbf{p}'_n$ noting that $n+1$ becomes 1 for a closed chain. Combining these expressions together gives

$$\mathbf{T}_{12}\mathbf{T}_{23}\mathbf{T}_{34}...\mathbf{T}_{n1} = \mathbf{I}, \tag{2.17}$$

where $\mathbf{I}$ is a 4×4 unit matrix. Eq. (2.17) can be used to derive the closure equations of the closed kinematic chain.

The matrix method, like any of the other kinematic methods, takes both the topology and the geometry of a linkage into account. It can be used to determine the mobility of a linkage. If Eq. (2.17) has only one or a limited number of solutions, the closed chain is in fact locked. When one of the kinematic variables can change freely while the others are found to be dependent upon it algebraically by the equations in Eq. (2.17), the linkage has mobility one. If two free kinematic variables exist, the linkage will have mobility two. The number of mobility of linkages is equal to the number of free kinematic variables in the closure equations. Meanwhile, other kinematic properties of the linkages, e.g. motion trajectories, can also be obtained from the closure equations.

The matrix method is also applicable to planar linkages. In a planar mechanism shown in Figure 2.9, all of the z axes are parallel and both

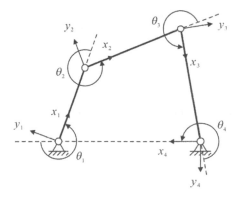

Figure 2.9 A planar linkage with a local coordinate system for each link.

twists and offsets are zero. The homogeneous coordinates can then be replaced by three dimensional vectors

$$\mathbf{p}'_i = \begin{bmatrix} 1 \\ x_i \\ y_i \end{bmatrix}, \tag{2.18}$$

and the transfer matrix becomes

$$\mathbf{T}_{i(i+1)} = \begin{bmatrix} 1 & 0 & 0 \\ -a_{i(i+1)} & \cos\theta_i & \sin\theta_i \\ 0 & -\sin\theta_i & \cos\theta_i \end{bmatrix}. \tag{2.19}$$

We still have

$$\mathbf{T}_{12}\mathbf{T}_{23}\mathbf{T}_{34}\ldots\mathbf{T}_{n1} = \mathbf{I}. \tag{2.20}$$

where $\mathbf{I}$ is now a 3×3 unit matrix.

For spherical mechanisms the lengths of all the links and offsets are always zero due to the fact that all the joint axes intersect at the concurrency point. And the links are presented by the twist between two joints connected to this link. Therefore, there is only rotational transformation and no translation transformation. Then the transfer matrix becomes

$$\mathbf{T}_{i(i+1)} = \begin{bmatrix} 1 & 0 & 0 & 0 \\ 0 & \cos\theta_i & \sin\theta_i & 0 \\ 0 & -\cos\alpha_{i(i+1)}\sin\theta_i & \cos\alpha_{i(i+1)}\cos\theta_i & \sin\alpha_{i(i+1)} \\ 0 & \sin\alpha_{i(i+1)}\sin\theta_i & -\sin\alpha_{i(i+1)}\cos\theta_i & \cos\alpha_{i(i+1)} \end{bmatrix}$$

For a single closed spherical linkage, we still have

$$\mathbf{T}_{12}\mathbf{T}_{23}\mathbf{T}_{34}\ldots\mathbf{T}_{n1} = \mathbf{I},$$

or

$$\mathbf{Q}_{12}\mathbf{Q}_{23}\mathbf{Q}_{34}\ldots\mathbf{Q}_{n1} = \mathbf{I}, \tag{2.21}$$

where $\mathbf{Q}$s are given in Eq. (2.10) and $\mathbf{I}$ is now a 3×3 unit matrix.

2.3 Overconstrained linkages

2.3.1 Introduction

We have shown examples that the Kutzbach criterion may count idle degrees of freedom in some mechanisms. It may also give a misleading result, which is smaller than the real degrees of freedom of the mechanisms, by disregarding the geometry of an assembly. Figure 2.10(a) is a planar 4R linkage. According to the Kutzbach criterion (2.2),

$$m = 3(4-4-1)+(4\times1) = 1.$$

If an additional rod EF is added into the linkage as Figure 2.10(b), Eq. (2.2) yields

$$m = 3(5-6-1)+(6\times1) = 0.$$

However, if AB=CD and EF=AD=BC, ABCD and AEFD are parallelograms, closure condition Eq. (2.5) is automatically met. The linkage has mobility one because of the special geometry it has.

It is important to realise that having a value less than one from the Kutzbach criterion does not automatically imply that the mechanism is a conventional structure. A mechanism can have a full range of mobility even though the Kutzbach criterion indicates otherwise. This type of mechanism is called the *overconstrained mechanism*. The existence of mobility is due to special geometry conditions that are known as the *overconstrained conditions*.

For a spatial closed chain where only lower pair joints are involved and each joint has one degree of freedom, the Kutzbach criterion becomes

$$m = 6n - 5j - 6 \tag{2.22}$$

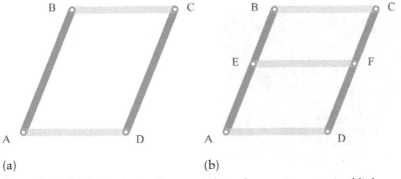

(a)   (b)

Figure 2.10 (a) A four-bar linkage and (b) a planar overconstrained linkage.

For such a chain to have mobility one ($m=1$) seven links ($n=7$) and seven joints ($j=7$) are needed. Thus, any kinematic chains with fewer links and joints are to be either immobile or overconstrained.

Closed chains with seven links are less suitable as building blocks for motion structures just like that heptagons are hardly used in surface tiling. Hence, overconstrained spatial closed chains have been found to be particularly useful in forming large motion structures by tessellation. Before embarking on the task of building motion structures let us first survey all of the existing overconstrained linkages with emphasis on spatial linkages for most of the planar overconstrained linkages are variations of the four-bar linkage, such as the one shown in Figure 2.10(b).

2.3.2 Early examples

The first published research on overconstrained mechanisms can be traced back to Sarrus (1853) when he reported a six-bar mechanism capable of rectilinear motion, Figure 2.11. More overconstrained mechanisms followed in the next half a century. Unfortunately most of the overconstrained mechanisms have rarely been used in real industrial applications because of the development of gears, cams and other means of transmission. But there are two exceptions, namely the double-Hooke's-joint linkage and the Schatz linkage (Phillips, 1990; Baker, 2002; Lee and Dai, 2003). The former has been widely applied as a transmission coupling, whereas the latter led to the Turbula machine for mixing fluids and powders.

2.3.3 The Bennett linkage

A chain of two or three links connected by the same number of revolute joints is found to be either a rigid structure or trivial mechanism when all three revolute axes are coplanar and intersect at a single point (Phillips, 1990). The minimum number of links used for construction of a nontrivial mobile chain with revolute joints is four, and the Bennett linkage (Bennett,

Figure 2.11 The Sarrus linkage.

1903) is the only 4R overconstrained linkage having the axes of four revolute joints neither parallel nor concurrent (Waldron, 1968; Savage, 1972; Baker, 1975), see Figure 2.12. Figure 2.13 shows one of the original models made by Bennett. This linkage was also found independently by Borel (Bennett, 1914). Bennett (1914) identified the conditions for the linkage to have a single degree of mobility as follows.

a   Two alternate links have the same length and the same twist, i.e.

$$a_{12} = a_{34} = a, \tag{2.23a}$$

$$a_{23} = a_{41} = b, \tag{2.23b}$$

$$\alpha_{12} = \alpha_{34} = \alpha, \tag{2.23c}$$

$$\alpha_{23} = \alpha_{41} = \beta. \tag{2.23d}$$

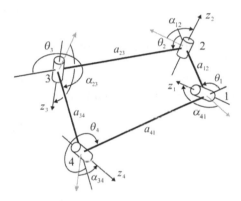

Figure 2.12 The Bennett linkage.

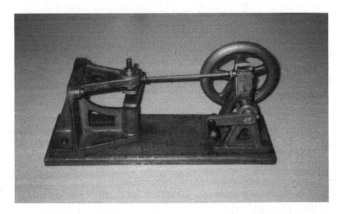

Figure 2.13 The original model of the Bennett linkage made by Bennett.

b   Lengths and twists should satisfy the condition

$$\frac{\sin \alpha}{a} = \frac{\sin \beta}{b}.$$ (2.24)

c   Offsets are zero, i.e.

$$R_i = 0 \ (i = 1, 2, 3 \text{ and } 4).$$ (2.25)

The revolute variables are $\theta_i (i = 1, 2, 3 \text{ and } 4)$. The closure equations of the linkage can be derived by applying the matrix method with the Denavit and Hartenberg notation.

From (2.17), we have

$$\mathbf{T}_{12} \mathbf{T}_{23} \mathbf{T}_{34} \mathbf{T}_{41} = \mathbf{I},$$

or

$$\mathbf{T}_{12} \mathbf{T}_{23} = \mathbf{T}_{14} \mathbf{T}_{43},$$ (2.26)

where 4×4 matrices $\mathbf{T}_{12}$, $\mathbf{T}_{23}$, $\mathbf{T}_{34}$ as well $\mathbf{T}_{41}$ are given by Eqs (2.14) and (2.16), respectively.

Eq. (2.26) contains a total of twelve equations and four identities. Among them are $\cos \theta_2 = \cos \theta_4$ and $\sin \theta_2 = -\sin \theta_4$ which lead to

$$\theta_2 + \theta_4 = 2\pi.$$ (2.27a)

Similarly,

$$\theta_1 + \theta_3 = 2\pi.$$ (2.27b)

The other one is

$$\sin \alpha \cos \beta \cos \theta_2 + \cos \alpha \sin \beta = \sin \beta \sin \theta_3 \sin \theta_4 - \cos \alpha \sin \beta \cos \theta_3 \cos \theta_4$$

$$- \sin \alpha \cos \beta \cos \theta_3,$$

which can be rewritten as

$$\tan \frac{\theta_1}{2} \tan \frac{\theta_2}{2} = \frac{\sin \dfrac{\beta + \alpha}{2}}{\sin \dfrac{\beta - \alpha}{2}}.$$ (2.27c)

Eq. (2.27) are the closure equations of the Bennett linkage. Among all of the revolute variable θs, only one is independent and the rest can be worked out from the closure equations. Consequently the linkage has a single mobility (Baker, 1979).

Bennett (1914) also identified some special cases.

a   An equilateral linkage is obtained if $\alpha + \beta = \pi$ and $a = b$. Eq. (2.27c) then becomes

$$\tan \frac{\theta_1}{2} \tan \frac{\theta_2}{2} = \frac{1}{\cos \alpha}. \tag{2.28}$$

b   If $\alpha = \beta$ and $a = b$, the four links are congruent. The motion is discontinuous: $\theta_1 = \pi$ allows any value for θ_2 and $\theta_2 = \pi$ allows any value for θ_1.

c   If $\alpha = \beta = 0$, the linkage is a planar four-bar linkage.

d   If $\alpha = 0$ and $\beta = \pi$, the linkage becomes a 2D parallelogram.

e   The linkage becomes a spherical 4R linkage if $a = b = 0$ (Phillips, 1990).

2.3.4 Linkages derived from the Bennett linkage

Attempts have also been made to build 5R or 6R three dimensional linkages based on the Bennett linkage. Most of the work was concentrated on building new mobile chains with fewer than seven links rather than exploring the possibility of constructing large motion assemblies with the only exception of Baker and Hu's (1986) unsuccessful attempt to connect two Bennett linkages, which we shall discuss in Chapter 5.

Goldberg (1943) arrived at the Goldberg 5R linkage by combining a pair of Bennett linkages in such a way that a link common to both was removed and a pair of adjacent links were rigidly attached to each other. The techniques he developed can be summarised as the *summation* of two Bennett loops to produce a 5R linkage, Figure 2.14(a), or the *subtraction* of a primary composite loop from another Bennett chain to form a syncopated linkage, Figure 2.14(b).

Prior to Goldberg, Myard (1931) produced an overconstrained 5R linkage as shown in Figure 2.15. It is a plane-symmetric 5R and has later been reclassified as a special case of the Goldberg 5R linkage, for which the two 'rectangular' Bennett chains with one pair of twists being $\pi/2$, are symmetrically disposed and subsequently combined. Two Bennett linkages are mirror images of each other where the mirror is coincident with the plane of symmetry of the resultant linkage (Baker, 1979). The conditions on its geometric parameters are as follows.

$$a_{34} = 0 \,, \; a_{12} = a_{51} \,, \; a_{23} = a_{45}$$

$$\alpha_{23} = \alpha_{45} = \frac{\pi}{2} \,, \; \alpha_{51} = \pi - \alpha_{12} \,, \; \alpha_{34} = \pi - 2\alpha_{12} \tag{2.29}$$

$$R_3 = R_4 = 0.$$

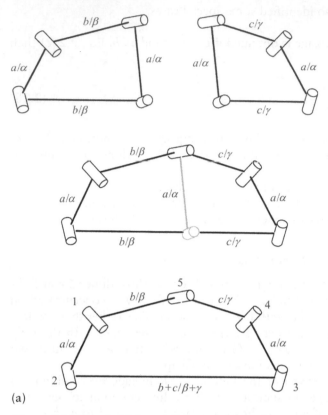

(a)

Figure 2.14 The Goldberg 5R linkages obtained by (a) summation; (b) subtraction.

An extended Myard linkage has been proposed by Chen and You (2008a) obtained by combining two Bennett linkages as Myard did in the creation of his 5R linkage. Unlike the original Myard linkage, the angle of twists in the extended Myard linkage, α_{23} and α_{45} are not necessary to be $\pi/2$ but the sum of these two twists remains to be π. And the two general Bennett linkages that have two equal lengths and one identical twist and one complimentary twist. As a result, the extended Myard linkage does not have plane symmetry. Similar to the Myard linkage, the extended one is also a special case of the generalised Goldberg 5R linkage.

Goldberg (1943) also reported a family of 6R linkages which were later named after him. Similar to the Goldberg 5R linkage, the Goldberg 6R linkages are also produced by combining Bennett linkages. There are a total of four types of Goldberg 6R linkages, see Figure 2.16. The first Goldberg 6R linkage is formed by arranging three Bennett linkages in series. The first two Bennett linkages have a link in common, and the opposite link of one of them is common with a link of a third Bennett linkage. The second Goldberg 6R

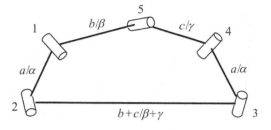

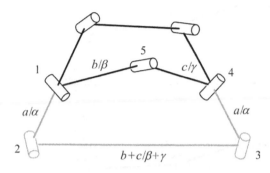

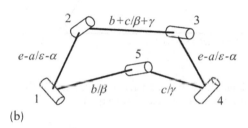

(b)

Figure 2.14 continued.

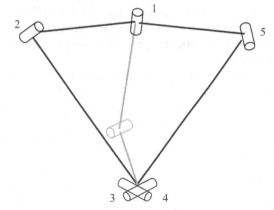

Figure 2.15 The Myard linkage.

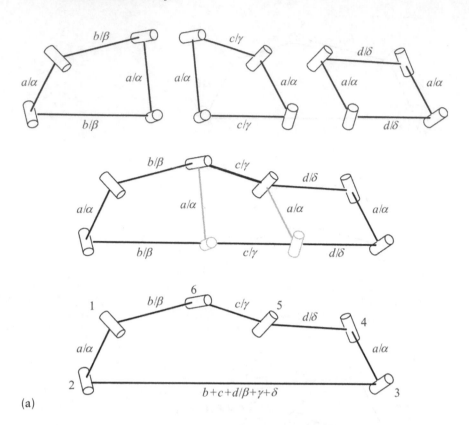

(a)

Figure 2.16 The Goldberg 6R linkages and their construction: (a) series; (b) syncopated; (b) syncopated; (c) L-shaped and (d) syncopated of L-shaped.

linkage is the *subtraction* of the first Goldberg 6R linkage from another Bennett linkage resulting in a *syncopated* linkage. The third Goldberg 6R linkage is built by an L-shaped arrangement of three Bennett linkages. The last of the four types is produced by subtracting the Goldberg L-shaped 6R linkage from another Bennett linkage.

Other 6R overconstrained linkages based on the Bennett linkage also exist, among which the most common one is the double-Hooke's-joint linkage (Baker, 2002), which we briefly touched upon earlier. It is a 6R linkage obtained from two spherical 4R linkages. This linkage is in fact a special case of the Bennett 6R hybrid linkage, which is a combination of two spherical 4R linkages. Two further special forms are the Bennett plano-spherical hybrid linkage, Figure 2.17, and the Sarrus linkage, Figure 2.11. Wohlhart (1991) described a 6R overconstrained linkage, the synthesis of which was achieved by coalescing two appropriate generalised Goldberg 5R linkages and removing the two common links, Figure 2.18.

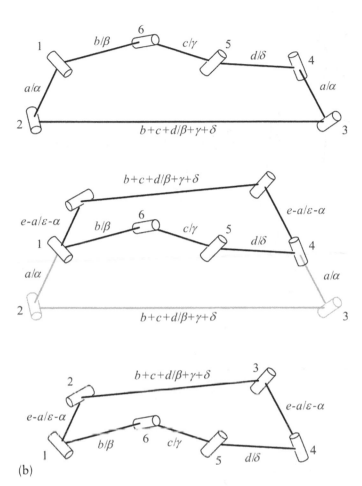

(b)

Figure 2.16 continued.

Chen and You (2007b) reported two types of 6R linkages, constructed from combination of two Goldberg 5R linkages in a matter similar to what Wohlhart did. If Wohlhart's combination is referred to as a *face-to-face* combination of two Goldberg 5R linkages, Chen and You's could be regarded as *back-to-back* as demonstrated in Figure 2.19. Mavroidis and Roth (1994) discovered a new overconstrained 6R linkage, the Bennett-joint 6R linkage, as a by-product of their effort to develop a systematic method to deal with overconstrained linkages. And Dietmaier 6R linkage was discovered with the aid of a numerical method (Dietmaier, 1995). More generally, Waldron (1968) drew attention to a class of mobile six-bar linkages with only lower pairs which included helical, cylinder and prism joints in addition to revolute joints. He suggested that any two single-loop linkages with a single degree of freedom could be arranged in a

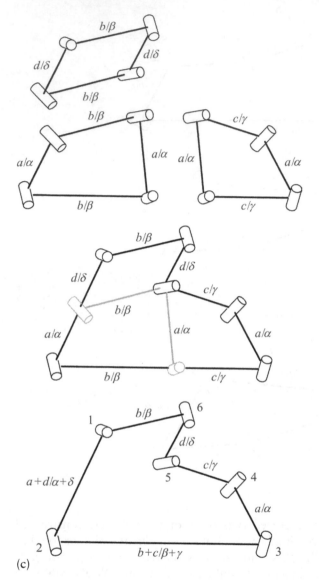

(c)

Figure 2.16 continued.

space to make them share a common axis, and, after this common joint was removed, the resulting linkage would certainly remain mobile. The condition for the resultant linkage to have mobility one is that the equivalent screw systems of the original linkages shall intersect only in the screw axis of the common joint when they are placed. Waldron listed all six-bar linkages, which can be formed from two four-bar linkages with lower joints. It is obvious that the double-Hooke's-joint linkage and the Bennett 6R linkages that we discussed above belong to this linkage family.

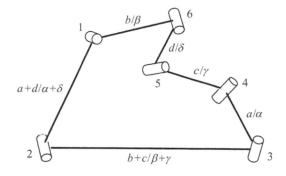

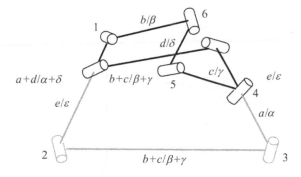

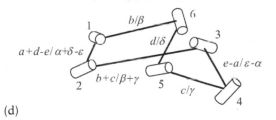

(d)

Figure 2.16 continued.

2.3.5 The Bricard linkages

Among six-bar overconstrained mechanisms with revolute joints, the most remarkable are the Bricard linkages. Bricard reported three different types of mobile 6R linkages when he was investigating octahedra (Bricard, 1897) and later he added three more types, namely the line-symmetric, the plane-symmetric and the trihedral linkages (Bricard, 1927). All of the six types and their associated geometrical conditions are summarised as follows (Baker, 1980; Phillips, 1990).

Figure 2.17 The Bennett plano-spherical hybrid linkage.

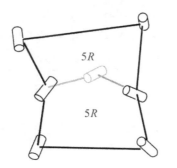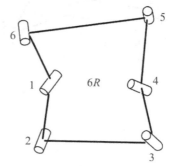

Figure 2.18 Wohlhart's double-Goldberg linkage.

a   The general line-symmetric case

$$a_{12} = a_{45} \,,\; a_{23} = a_{56} \,,\; a_{34} = a_{61};$$

$$\alpha_{12} = \alpha_{45} \,,\; \alpha_{23} = \alpha_{56} \,,\; \alpha_{34} = \alpha_{61}; \tag{2.30}$$

$$R_1 = R_4 \,,\; R_2 = R_5 \,,\; R_3 = R_6.$$

b   The general plane-symmetric case

$$a_{12} = a_{61} \,,\; a_{23} = a_{56} \,,\; a_{34} = a_{45};$$

$$\alpha_{12} + \alpha_{61} = 2\pi \,,\; \alpha_{23} + \alpha_{56} = 2\pi \,,\; \alpha_{34} + \alpha_{45} = 2\pi; \tag{2.31}$$

$$R_1 = R_4 = 0 \,,\; R_2 = -R_6 \,,\; R_3 = -R_5.$$

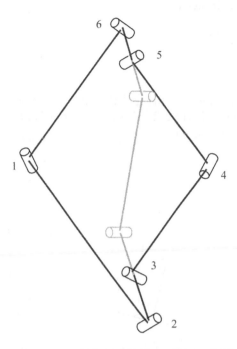

Figure 2.19 Chen and You's double-Goldberg linkage.

c   The trihedral case

$$a_{12}^2 + a_{34}^2 + a_{56}^2 = a_{23}^2 + a_{45}^2 + a_{61}^2;$$

$$\alpha_{12} = \alpha_{34} = \alpha_{56} = \frac{\pi}{2}, \ \alpha_{23} = \alpha_{45} = \alpha_{61} = \frac{3\pi}{2}; \tag{2.32}$$

$$R_i = 0 \ (i = 1, 2, \cdots, 6).$$

d   The line-symmetric octahedral case

$$a_{12} = a_{23} = a_{34} = a_{45} = a_{56} = a_{61} = 0; \tag{2.33}$$

$$R_1 + R_4 = R_2 + R_5 = R_3 + R_6 = 0.$$

e   The plane-symmetric octahedral case

$$a_{12} = a_{23} = a_{34} = a_{45} = a_{56} = a_{61} = 0;$$

$$R_4 = -R_1, \ R_2 = -R_1 \frac{\sin \alpha_{34}}{\sin(\alpha_{12} + \alpha_{34})}, \ R_5 = R_1 \frac{\sin \alpha_{61}}{\sin(\alpha_{45} + \alpha_{61})}, \tag{2.34}$$

$$R_3 = R_1 \frac{\sin \alpha_{12}}{\sin(\alpha_{12} + \alpha_{34})}, \ R_6 = -R_1 \frac{\sin \alpha_{45}}{\sin(\alpha_{45} + \alpha_{61})}.$$

f   The doubly collapsible octahedral case

$$a_{12} = a_{23} = a_{34} = a_{45} = a_{56} = a_{61} = 0;$$ (2.35)

$$R_1 R_3 R_5 + R_2 R_4 R_6 = 0.$$

Motion sequence of the cases (a) and (b) Bricard linkages are shown in Figures 2.20 and 2.21, respectively. Figure 2.22 shows two models of cases (c) and (d).

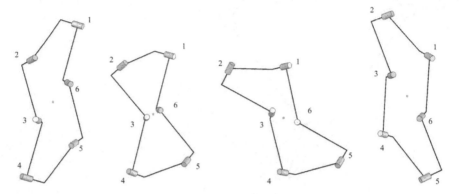

Figure 2.20 Motion sequence of a line-symmetric Bricard linkage.

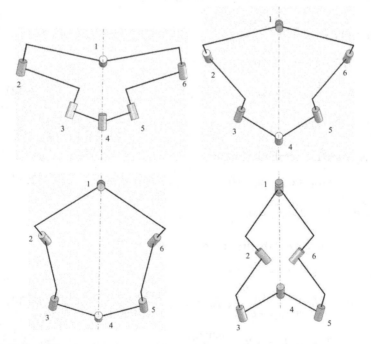

Figure 2.21 Motion sequence of a plane-symmetric Bricard linkage.

(a)   (b)

Figure 2.22 More Bricard linkages: (a) trihedral and (b) line-symmetric octahedral cases.

The Bricard linkages also appear in other objects or linkages. A linkage popularly known as the kaleidocycle is one example. A kaleidocycle is a three dimensional ring made from a chain of identical tetrahedra, each of which is linked to an adjoining one along an edge. The ring can be turned through its centre continuously. At least six tetrahedra are required in order to form a mobile closed ring (Schattschneider and Walker, 1977). In fact when the number of tetrahedra is six, the loop becomes a trihedral case of the Bricard linkage. A model is shown in Figure 2.23, which is constructed from a toy commonly known as Flexistar 6.[1] Its geometrical properties are as follows.

$$a_{12} = a_{23} = a_{34} = a_{45} = a_{56} = a_{61},$$

$$\alpha_{12} = \alpha_{34} = \alpha_{56} = \frac{\pi}{2}, \ \alpha_{23} = \alpha_{45} = \alpha_{61} = \frac{3\pi}{2}, \tag{2.36}$$

$$R_i = 0 \ (i = 1, 2, \ldots, 6).$$

The other example is the Altmann 6R linkage (1954), which also turned out to be a special case of the line-symmetric and trihedral case Bricard linkage. Moreover, the Schatz linkage, reported and patented by Schatz, was derived from a special trihedral Bricard linkage (Phillips, 1990). The new 6R linkage reported by Wohlhart (1987) can be regarded as a generalisation of the Bricard trihedral 6R linkage.

Figure 2.23 Motion sequence of a kaleidocycle toy made of six tetrahedra.

2.3.6 Summary

There are a total of sixteen types of spatial overconstrained linkages with revolution joints. A summary is given in Table 2.3. Only two of these linkages, namely the Bennett linkage and the Bricard linkage, can be regarded as completely independent, whereas the rest are combinations or derivatives of the two linkages.

2.4 Mechanisms and structures

When only a unique solution exists for the closure equations (2.17), the closed chain has zero mobility and is in fact a conventional structure because none of n kinematic variables can change freely. The tetrahedron given in Figure 2.2 is an example.

Structural engineers are also interested in finding out whether an assembly is a structure or a mechanism. A criterion known as the Maxwell's rule is used to determine whether an assembly is a structure or mechanism (Calladine, 1978; Pellegrino and Calladine, 1986; Calladine and Pellegrino, 1991). It stipulates that a truss in space with j spherical joints requires in general n bars and r supports to render it *stiff* (or rigid) where $n+r=3j$; when $n+r>3j$, the structure would have *redundant* members and when $n+r<3j$, the assembly could be a mechanism. For trusses without ground support, r is taken as 6 to remove the rigid body motions associated to the trusses as a whole.

Table 2.3 Spatial overconstrained linkages with only revolute joints and their dependent linkages.

Number of links	Linkages	Dependent linkages
4	Bennett linkage	—
5	Goldberg 5R linkage	Bennett linkage
5	Myard linkage	Bennett linkage
6	Altmann linkage	Bricard linkage
6	Bennett 6R hybrid linkage	Bennett linkage
6	Bennett-joint 6R linkage	Bennett linkage
6	Bricard linkages	—
6	Dietmaier 6R linkage	Bennett linkage
6	Double-Hooke's-joint linkage	Bennett linkage
6	Goldberg 6R linkage	Bennett linkage
6	Sarrus linkage	Bennett linkage
6	Schatz linkage	Bricard linkage
6	Waldron hybrid linkages	Four-bar linkage with lower joints
6	Wohlhart 6R linkage	Bricard linkage
6	Wohlhart double-Goldberg linkage	Bennett linkage
6	Chen & You double-Goldberg linkage	Bennett linkage

Applying the Maxwell's rule to the tetrahedron shown in Figure 2.2 we have $n = 6$, $r = 6$ and $j = 4$. The tetrahedron truss is therefore a stiff structure because $3j - (n + r) = 0$. Unlike the Kutzbach criterion, the Maxwell's rule can disregard the idle degrees of freedom in this instance.

Similar to the Kutzbach criterion, the Maxwell's rule does not take the geometry of the assembly into consideration. It is no surprise that it can also produce misleading results. To avoid that, a more thorough approach has to be taken.

Consider a truss in space with n members, r supports and j spherical joints. It is subjected to $3j$ arbitrary independent components of load applied at its joints, and let these be denoted by vector f. The $3j$ equations of equilibrium relate the load components to the n member forces and the r reactions; denote by the vector t these unknown force variables. The equilibrium equations for the original, undeformed frame may be written as

$$\mathbf{Ht} = \mathbf{f}. \tag{2.37}$$

The equilibrium matrix H has $3j$ rows and $n + r$ columns.

Corresponding to f is the vector d of (small) nodal displacements; and corresponding to t is the elongation vector e considering of n bar extensions and r ground displacements, all small. These quantities are related, for small displacements, by the kinematic relations

$$\mathbf{Cd} = \mathbf{e}. \tag{2.38}$$

The compatibility matrix C thus has $n + r$ rows and $3j$ columns.

Vectors e and t are related by the constitutive laws of the material and sectional dimensions.

The principle of virtual work states that $\mathbf{f}^T\mathbf{d} = \mathbf{t}^T\mathbf{e}$, leading to $\mathbf{C} = \mathbf{H}^T$.

Eqs (2.37) and (2.38) give rise to a number of possible scenarios depending on the rank of H, r_H.

First, if $3j = n + r$ and r_H is equal to $3j$, matrix H is a non-singular square matrix and the following solution can be obtained

$$\mathbf{t} = \mathbf{H}^{-1}\mathbf{f}. \tag{2.39}$$

Such an assembly is classified as *statically determinate* since the internal forces in the assembly can determine uniquely by application of the static equilibrium alone. In particular, if $\mathbf{f} = 0$, $\mathbf{t} = 0$, and consequently members carry no forces, the matrix C ($= \mathbf{H}^T$) has the same rank as H and is also non-singular, so Eq. (2.38) gives

$$\mathbf{d} = \mathbf{C}^{-1}\mathbf{e}. \tag{2.40}$$

For $\mathbf{e} = 0$ (the bars would have no elongations as they carry no forces), $\mathbf{d} = 0$. The frame is therefore stiff.

Next consider that matrix $\mathbf{H}$ of a certain frame is neither square $(3j \neq n + r)$ nor full rank $(r_H < 3j$ or $r_H < n + r)$, then let

$$s = n + r - r_H, \tag{2.41a}$$

$$m = 3j - r_H. \tag{2.41b}$$

If $s > 0$, then the columns of $\mathbf{H}$ are linearly dependent, and the equilibrium Eq. (2.39) will have non-zero solutions for t even if f = 0. The assembly can have a total of s sets of linearly independent internal forces without the external loading. It is therefore regarded as being *statically indeterminate* because the equations of equilibrium alone are insufficient to uniquely determine the member forces. The forces are known as the *states of self-stress*.

When $m > 0$ the columns of $\mathbf{C}$ are linearly dependent, and similarly the compatibility equation can have non-zero solutions for d even though e = 0. The assembly can have m sets of linearly independent displacements. It is known as being *kinematically indeterminate* since the displacements of the joints cannot be uniquely determined by the lengths of the members.

The static and kinematic characteristics of an assembly can be given by the pair s and m, both of which can either be greater than or equal to zero. The possibilities can be grouped into a total of four categories.

a   the assembly is both statically and kinematically determinate. It has neither state of self-stress nor mechanism ($s = 0$, $m = 0$);
b   the assembly is statically determinate and kinematically indeterminate frame. It has no state of self-stress but is a mechanism with mobility m ($s = 0$, $m > 0$);
c   the assembly is statically indeterminate and kinematically determinate. It has states of self-stress and is stiff ($s > 0$, $m = 0$); and
d   the assembly is both statically and kinematically indeterminate. It has states of self-stress but at the same time it is a mechanism with m mobilities ($s > 0$, $m > 0$).

The same classification is applicable to assemblies other than trusses. The common linkages surveyed in previous sections belong to (b) whereas the overconstrained linkages belong to (d) which become statically indeterminate structures once their motion is locked. The existence of self-stress can be used to detect the mobility in overconstrained linkages.

Readers should be aware that the above linear algebraic analysis, set up only for the initial geometrical configuration, has its own limitations despite taking both geometry and topology of the assembly into account. The displacements may be infinitesimal, i.e. the assembly will tighten up after a small displacement, instead of full cycle mobility (or being truly mobile). The advanced materials on the topic can be found in Calladine (1978), Pellegrino and Calladine (1986) and Tarnai (1984, 2001).

3 Planar double chain linkages

DOI: 10.1201/9781482266610-3

3.1 Scissor-like elements and their assemblies

Many readers were attracted to deployable structures by a toy known as the Hoberman sphere, Figure 3.1(a). The toy is a one-degree-of-freedom mechanism which resembles a sphere but is capable of expanding up to a

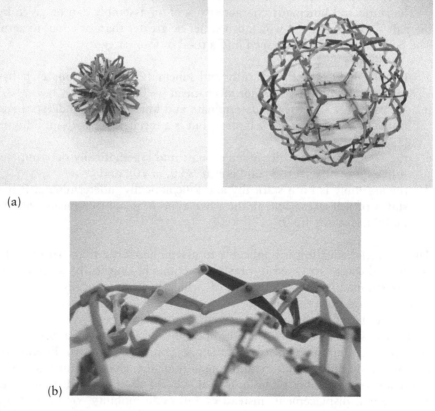

(a)

(b)

Figure 3.1 (a) Expansion sequence of the Hoberman sphere and (b) its structural detail.

few times of its packaged size. The prime component of the toy is the scissor-like element (also referred to as a pantograph in some references), Figure 3.1(b), which consists of a pair of beams joined together by a pivot (revolute joints) so that free rotation of one beam relative to another about the axis of the pivot is allowed and any other relative motion of the rods is prevented. The scissor-like element is probably the simplest mechanism involving a revolute joint.

There are two types of scissor-like elements depending on the locations of the hinges and pivot on the beams that form the element. In the first type, Figure 3.2(a), referred to here as *the conventional scissor-like element*, the pivot and two end connectors, used for connection with neighbouring elements, are collinear. Therefore, the beams can be made from straight rods. In the second type however, the end hinges and pivot are non-collinear, Figure 3.2(b). Kinked beams or flat plates are therefore used, which gives the name: *the angulated scissor-like element*.

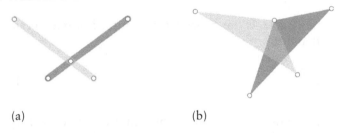

(a)   (b)

Figure 3.2 (a) The conventional and (b) angulated scissor-like elements.

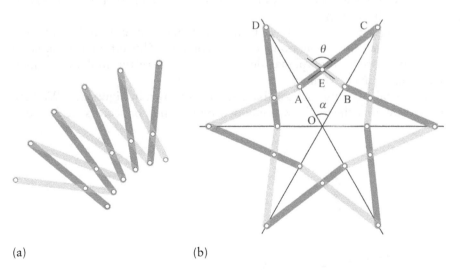

(a)   (b)

Figure 3.3 (a) A mobile chain made of conventional scissor-like elements that spans to a curved profile; (b) a ring of zero mobility, formed by six identical conventional scissor-like elements.

The conventional scissor-like element is quite versatile. A number of the elements can be placed in sequence to form a deployable assembly like that of the lazy tong shown in Figure 2.3. This type of assembly is also known as *double chain* because it appears like two interwoven individual chains. The double chain has mobility one, for the pivoting angle of one element acts as the input of its neighbour. Carefully designed double chain can even expand to a curved profile, Figure 3.3(a), but the double chain containing only conventional scissor-like elements should never be closed since the closure, i.e. the first element is connected with the last, will render it to a structure. To prove it, consider a ring of six ($n=6$) identical conventional scissor-like elements shown in Figure 3.3(b). Each of the elements is made from two identical straight beams pivoted together and it occupies a sector with a subtended central angle α. There must be

$$\alpha = \frac{2\pi}{n} = \frac{\pi}{3}.$$

Denote by θ the pivoting angle. It can be shown that there is a one to one relationship between α and θ:

$$\tan\frac{\alpha}{2} = \frac{CE-EB}{AC}\tan\frac{\theta}{2}. \tag{3.1}$$

Hence, θ cannot be altered once α is known, which indicates that the ring has zero mobility and is in fact a structure. In other words, the conventional scissor-like elements cannot be used to construct mobile planar closed double chains.

The angulated scissor-like element that Hoberman used in his sphere is different from the conventional one in geometry. This difference enables it to be used in forming planar double chains where the mobility of individual elements is retained.

A typical angulated scissor-like element is shown in Figure 3.4. We shall now show that the angle α, subtended by the end connectors A, B, C and

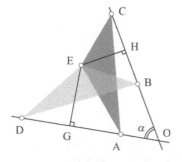

Figure 3.4 An angulated element made of two identical angulated beams.

D of a angulated scissor-like element, does not change when the angulated beams AEC and BED are rotated relative to one another if two beams are identical, i.e.

$$AE = DE, \ BE = CE^1 \tag{3.2}$$

and

$$\angle AEC = \angle DEB. \tag{3.3}$$

Let us draw two perpendiculars EG and EH on OD and OC, respectively. Considering deltoid OGEH, we have

$$\alpha = \pi - \angle GEH. \tag{3.4}$$

Because

$$\angle GEH = \angle AEG + \angle AEB + \angle BEH = \frac{\angle DEB - \angle AEB}{2} + \angle AEB$$

$$+ \frac{\angle AEC - \angle AEB}{2} = \frac{\angle DEB + \angle AEC}{2} = \angle AEC \text{ or } \angle DEB$$

substituting the above into Eq. (3.4) yields

$$\alpha = \pi - \angle AEC. \tag{3.5}$$

As $\angle AEC$ is predetermined, α is a constant regardless of the rotation between angulated beams AEC and BED.

This particular geometrical feature of the angulated scissor-like elements enables the construction of mobile plane loops. For a total of n angulated scissor-like elements, each of which has a subtended central angle α_i ($i = 1$, $2, \ldots, n$), to form a mobile double chain, there must be

$$\sum_{i=1}^{n} \alpha_i = 2\pi. \tag{3.6}$$

The Hoberman sphere has a number of predominant planar double chains composed of angulated scissor-like elements placed along the great circles of the sphere. For each double chain Eq. (3.6) holds.

Eq. (3.6) concerns angles only. It alone is not enough for the loop to be mobile. Figure 3.5(a) shows that a double chain of six angulated scissor-like elements, each consisting of a pair of identical beams that subtend a central angle of $\pi/3$. Eq. (3.6) is therefore satisfied and the lines linking two end connectors always remain parallel. However, there is a gap between the end connectors of the first and last elements. This gap may be bridged

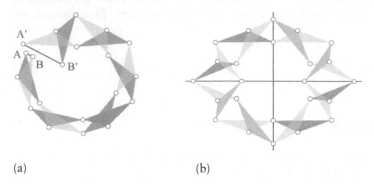

Figure 3.5 (a) A chain made from angulated elements, each subtends a constant central angle; (b) a mobile closed chain of angulated elements constructed using two-fold symmetry.

at a certain pivoting angle but it reappears if the angle varies. In other words, even if the connectors could be connected to form a loop at a particular pivoting angle, the resulted assembly will be immobile because a slight change of pivoting angle will result in reappearance of the gap.

This problem can be solved by making use of symmetry. Figure 3.5(b) is a chain made of the same pairs but in an order which is different from that in Figure 3.5(a). The chain has two-fold symmetry and sum of the central angles for each quarter is $\pi/2$ so that no split occurs when four quarters are joined together when the pivoting angle varies. Closure of the double chain becomes possible. Moreover, the closed double chain has mobility one.

In terms of practicality, symmetry is a good way to construct a mobile planar double chain of angulated scissor-like elements. However, other more general solutions exist which are obtained by considering the geometry of the entire assembly using the kinematic analysis tool that was introduced in Section 2.2.

3.2 Closed double chain

3.2.1 Background

The conquest of constructing mobile closed double chains dates back to over a century ago when Kempe (1878) first reported that under certain geometrical circumstances an assembly of two planar 4R linkages connected together by four additional hinges could become mobile. Six classes of such linkages known as the Kempe linkages were discovered, one of which is shown in Figure 3.6. Subsequently Darboux and Fontené provided further proofs of mobility for the Kempe linkage. A summary of their work can be found in Baker and Yu (1983).

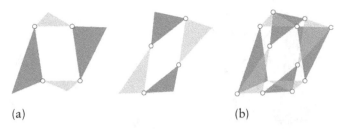

Figure 3.6 (a) Two four bar linkages and (b) a Kempe linkage formed by joining two four bar linkages.

The Kempe linkage displayed in Figure 3.6 can also be regarded as a closed double chain linkage consisting of *four* scissor-like elements. It complies with Eq. (3.6) even though α_i for each element may vary with pivoting angle whereas it is constant for the elements discussed in the previous section. Naturally one may well ponder whether similar mobile closed double chains exist should the number of pairs be other than four.

For a planar double chain linkage consisting of n elements it will have $3n$ hinges in total. According to the Kutzbach criterion for planar linkages consisting of only revolute joints Eq. (2.3), the degree of freedom of a closed double chain is

$$m = 3(2n - 3n - 1) + 3n = -3. \tag{3.7}$$

Hence, it is overconstrained. Should a chain become mobile, the existence of mobility is due to the particular geometry of the linkage.

To facilitate a more general discussion on double chain linkages, two types of angulated scissor-like elements, namely *the intersecting element* and *non-intersecting element*, are to be considered, both of which are shown in Figure 3.7(a). Although a large relative rotation can transform a non-intersecting element to an intersecting one as far as a single element is concerned, this kind of motion is physically prohibited when the elements are connected to form a closed double chain. Figures 3.7(b) and (c) show two closed double chain linkages made from intersecting and non-intersecting elements. From one of the end hinges drawing vectors continuously connecting two end hinges, it can be found that the vectors pass through all of the beams for a five-element double chain made from only intersecting elements, whereas they pass through only half of the beams for double chains made from only non-intersecting elements. Thus, it is necessary to consider the two cases separately. It turns out that their mobility conditions are different (Mao *et al.*, 2009).

Both the conventional and angulated scissor-like elements discussed in the previous section belong to the intersecting element. They are of a more practical type in forming engineering expandable structures.

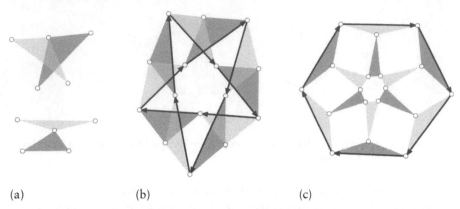

(a)   (b)   (c)

Figure 3.7 (a) Intersecting and non-intersecting elements; (b) and (c) closed double chains formed by intersecting and non-intersecting elements, respectively.

3.2.2 Double chains with intersecting elements

A double chain consisting of five intersecting elements prior to forming a closed loop is shown in Figure 3.8. Two beams in each element, Figure 3.8(a), can be represented by vectors $\mathbf{p}_i$, $\mathbf{p}_{i+1}$, $\mathbf{q}_i$ and $\mathbf{q}_{i+1}$, respectively ($i = 1, 2, \ldots, 5$. When the subscript becomes 6, it is replaced by 1). Denote by μ_i the inclination angle from $\mathbf{p}_i$ to $\mathbf{p}_{i+1}$ and by v_i the inclination angle from $\mathbf{q}_i$ to $\mathbf{q}_{i+1}$. Both of the angles range from $-\pi$ to π and they are positive clockwise.[2]

The chain is constructed subjected to the following restrictions:

a   within four neighbouring pieces, four hinges, two from each piece, form a parallelogram; and
b   at both ends, AB and EF are equal and parallel; so are BC and DE, see Figure 3.8(b).

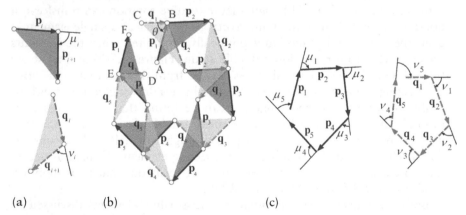

(a)   (b)   (c)

Figure 3.8 (a) Angulated beams forming an intersecting element; (b) a double chain prior to closure and (c) relations among inclination angles.

The above restrictions are referred to as the *loop parallelogram constraint* hereafter.

Under the loop parallelogram constraint, the inclination angles

$$\mu_1 + \mu_2 + \mu_3 + \mu_4 + \mu_5 = 2\pi, \tag{3.8a}$$

and

$$v_1 + v_2 + v_3 + v_4 + v_5 = 2\pi, \tag{3.8b}$$

as shown in Figure 3.8(c). Moreover, vectors

$$\overline{AC} = \overline{DF}. \tag{3.9}$$

because both AC and DF are equal to $\mathbf{p}_1 - \mathbf{q}_1$.

The condition for forming a closed loop is that the end connectors A and D should always meet when the mechanism is activated, Figure 3.9(a), and therefore,

$$\mathbf{p}_1 + \mathbf{q}_2 + \mathbf{p}_2 + \mathbf{q}_3 + \mathbf{p}_3 + \mathbf{q}_4 + \mathbf{p}_4 + \mathbf{q}_5 + \mathbf{p}_5 + \mathbf{q}_1 = 0. \tag{3.10}[3]$$

Now assume that the motion of the mechanism is determined by pivoting angle θ ($-\pi \leq \theta \leq \pi$). It is the angle between vector $\mathbf{q}_1$ and $\mathbf{p}_1$ and is positive clockwise as shown in Figure 3.9(a).[4] The vector equation (3.10) can be replaced by complex number notations with the real and imaginary axes parallel with and perpendicular to vector $\mathbf{q}_1$, respectively. There is

$$p_1 e^{j\theta} + q_2 e^{j(-v_1)} + p_2 e^{j(\theta - \mu_1)} + q_3 e^{j(-v_1 - v_2)} + p_3 e^{j(\theta - \mu_1 - \mu_2)} + q_4 e^{j(-v_1 - v_2 - v_3)} +$$
$$p_4 e^{j(\theta - \mu_1 - \mu_2 - \mu_3)} + q_5 e^{j(-v_1 - v_2 - v_3 - v_4)} + p_5 e^{j(\theta - \mu_1 - \mu_2 - \mu_3 - \mu_4)} + q_1 e^{j \cdot 0} = 0, \tag{3.11}$$

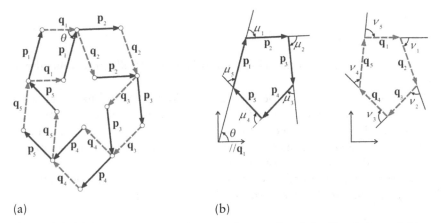

(a)   (b)

Figure 3.9 (a) Five intersecting elements form a closed double chain and (b) geometrical representation of its mobility conditions.

or

$$\left(p_1 + p_2 e^{j(-\mu_1)} + + p_3 e^{j(-\mu_1-\mu_2)} + p_4 e^{j(-\mu_1-\mu_2-\mu_3)} + p_5 e^{j(-\mu_1-\mu_2-\mu_3-\mu_4)} \right) e^{j\theta} +$$

$$\left(q_1 e^{j\times 0} + q_2 e^{j(-\nu_1)} + q_3 e^{j(-\nu_1-\nu_2)} + q_4 e^{j(-\nu_1-\nu_2-\nu_3)} + q_5 e^{j(-\nu_1-\nu_2-\nu_3-\nu_4)} \right) = 0. \quad (3.12)$$

To have a mobile linkage, the above equation must be satisfied whatever θ is, which leads to

$$p_1 + p_2 e^{j(-\mu_1)} + + p_3 e^{j(-\mu_1-\mu_2)} + p_4 e^{j(-\mu_1-\mu_2-\mu_3)} + p_5 e^{j(-\mu_1-\mu_2-\mu_3-\mu_4)} = 0 \quad (3.13)$$

and

$$q_1 e^{j\times 0} + q_2 e^{j(-\nu_1)} + q_3 e^{j(-\nu_1-\nu_2)} + q_4 e^{j(-\nu_1-\nu_2-\nu_3)} + q_5 e^{j(-\nu_1-\nu_2-\nu_3-\nu_4)} = 0. \quad (3.14)$$

Eqs (3.13) and (3.14) are the additional mobility conditions.

A close inspection of Eq. (3.13) reveals that geometrically it is equivalent to vectors $\mathbf{p}_1$, $\mathbf{p}_2$,..., $\mathbf{p}_5$ forming a closed loop. Similarly, Eq. (3.14) suggests that vectors $\mathbf{q}_1$, $\mathbf{q}_2$,..., $\mathbf{q}_5$ also form a closed loop. Both cases are shown in Figure 3.9(b). Moreover, if vectors $\mathbf{p}$ and $\mathbf{q}$ form closed loops, change of the pivoting angle, θ, does not alter their relative positions. Only the vector loops as a whole will rotate by the same amount, see Figure 3.9(b).

Hence, we can conclude that the conditions that ensure the formation of a mobile closed double chain consisting of five intersecting pairs are as follows.

$$\mathbf{p}_1 + \mathbf{p}_2 + \mathbf{p}_3 + \mathbf{p}_4 + \mathbf{p}_5 = 0, \quad (3.15a)$$

and

$$\mathbf{q}_1 + \mathbf{q}_2 + \mathbf{q}_3 + \mathbf{q}_4 + \mathbf{q}_5 = 0. \quad (3.15b)$$

The motion sequence of a model shown in Figure 3.10 illustrates the above proof.

The above derivation can be extended to double chain linkage consisting of n intersecting pairs by adding or reducing items in equations. The mobility condition for double chains composed of intersecting pairs under the loop parallelogram constraint is that two vector sums of edges of the pieces, defined as $\mathbf{p}$ and $\mathbf{q}$, must be zero.

Note that here the number of pairs, n, can be either *odd* or *even*. We shall explain the significance of this in the next section.

Figure 3.10 Expansion of a double chain with five angulated elements.

3.2.3 Double chains with non-intersecting elements

First consider a double chain linkage with four non-intersecting pairs as shown in Figure 3.11. The assembly is constructed so that it meets the loop parallelogram constraint. The loop parallelogram constraint gives

$$\mu_1 + \mu_2 + \mu_3 + \mu_4 = 2\pi, \tag{3.16a}$$

and

$$v_1 + v_2 + v_3 + v_4 = 2\pi. \tag{3.16b}$$

For the double chain shown in Figure 3.11 bringing A and D together will enable the double chain to be closed. Hence,

$$\mathbf{p}_1 + \mathbf{p}_2 + \mathbf{q}_2 + \mathbf{q}_3 + \mathbf{p}_3 + \mathbf{p}_4 + \mathbf{q}_4 + \mathbf{q}_1 = \mathbf{0}. \tag{3.17}$$

Introducing complex number notations,

$$p_1 e^{j\theta} + p_2 e^{j(\theta - \mu_1)} + q_2 e^{j(-v_1)} + q_3 e^{j(-v_1 - v_2)} + p_3 e^{j(\theta - \mu_1 - \mu_2)} + p_4 e^{j(\theta - \mu_1 - \mu_2 - \mu_3)} +$$

$$q_4 e^{j(-v_1 - v_2 - v_3)} + q_1 e^{j \times 0} = 0, \tag{3.18}$$

or

$$\left(p_1 + p_2 e^{j(-\mu_1)} + p_3 e^{j(-\mu_1 - \mu_2)} + p_4 e^{j(-\mu_1 - \mu_2 - \mu_3)} \right) e^{j\theta} + \left(q_1 e^{j \times 0} + q_2 e^{j(-v_1)} \right.$$

$$\left. + q_3 e^{j(-v_1 - v_2)} q_4 e^{j(-v_1 - v_2 - v_3)} \right) = 0. \tag{3.19}$$

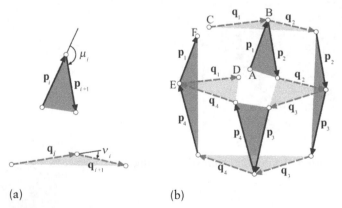

(a)   (b)

Figure 3.11 (a) Angulated beams forming a non-intersecting element and (b) a double chain linkage with four non-intersecting pairs prior to closure.

In order for the equation to hold for any θ, both the first and second items must be zero. which, together with Eq. (3.16), have been found to be equivalent to vectors $\mathbf{p}_1$, $\mathbf{p}_2$, $\mathbf{p}_3$, $\mathbf{p}_4$ and $\mathbf{q}_1$, $\mathbf{q}_2$, $\mathbf{q}_3$, $\mathbf{q}_4$ forming closed loops, respectively, i.e.

$$\mathbf{p}_1 + \mathbf{p}_2 + \mathbf{p}_3 + \mathbf{p}_4 = \mathbf{0}, \qquad (3.20\text{a})$$

and

$$\mathbf{q}_1 + \mathbf{q}_2 + \mathbf{q}_3 + \mathbf{q}_4 = \mathbf{0}, \qquad (3.20\text{b})$$

as shown in Figure 3.12. Alteration of rotation angle θ only causes the rotation of the closed vector loops as a whole.

This proof can be easily extended to double chain with any *even* number of non-intersecting pairs.

Although the appearance of Eqs (3.20a) and (3.20b) is similar to what we obtained for the double chain with intersecting pairs, they are in fact different. This is due to the way the pieces were arranged under the loop parallelogram constraint, see Figure 3.12(a). Here $\mathbf{p}_1$, $\mathbf{p}_2$, $\mathbf{p}_3$ and $\mathbf{p}_4$ represent the lengths of opposite pieces. So do $\mathbf{q}_1$, $\mathbf{q}_2$, $\mathbf{q}_3$ and $\mathbf{q}_4$.

Now consider a double chain consisting of five non-intersecting pairs, see Figure 3.13(a). Note that, unlike the previous double chain with four non-intersecting elements, the element on the left consists of a piece with edge lengths of $\mathbf{p}_5$ and $\mathbf{q}_1$ whereas the other piece has edge lengths of $\mathbf{q}_5$ and $\mathbf{p}_1$ in order to preserve the loop parallelogram constraint. The angles sustained by $\mathbf{p}_5$ and $\mathbf{q}_1$, and by $\mathbf{q}_5$ and $\mathbf{p}_1$, are represented by μ_5 and ν_5, respectively.

Plotting all of the vectors ps and then qs, under the parallelogram constraint, see Figure 3.13(b), we obtain

$$\mu_1 + \mu_2 + \mu_3 + \mu_4 + \mu_5 + \theta = 2\pi, \qquad (3.21\text{a})$$

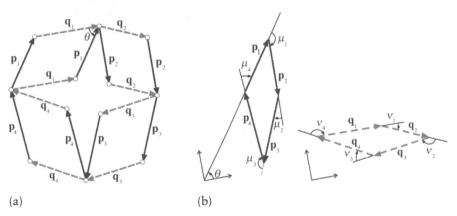

(a)   (b)

Figure 3.12 (a) Four non-intersecting elements form a closed double chain and (b) geometrical representation of its mobility conditions.

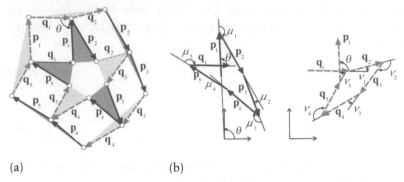

Figure 3.13 (a) A double chain linkage with five non-intersecting elements and (b) the vectors representing the beams and the pivoting angle.

and

$$v_1 + v_2 + v_3 + v_4 + v_5 - \theta = 2\pi, \tag{3.21b}$$

in which θ is the motion angle between vectors p_1 and q_1.

Since μs and vs are constants, obviously Eqs (3.21a) and (3.21b) cannot be maintained unless pivoting angle θ were constant. This simply shows that the assembly has no mobility.

The same conclusion can be drawn for all of the double chains with an *odd* number of non-intersecting elements.

In summary, the mobility condition for a double chain linkage mode from non-intersecting elements subjected to the loop parallelogram constraint is identical to those for the double chain linkage consisting of intersecting elements except that the number of pairs must be *even*.

This result was first obtained by Wohlhart (2000), but here we have used the vector method with complex number notation to prove them.

In the next section, we shall extend this approach to show that it is possible to construct double chain linkages with a mixture of intersecting and non-intersecting pairs.

3.2.4 General double chain linkages

Now examine a double chain linkage consisting of a total of five elements: three intersecting elements connected with two non-intersecting ones, as shown in Figure 3.14(a). The loop parallelogram constraint gives

$$\mu_1 + \mu_2 + \mu_3 + \mu_4 + \mu_5 = 2\pi, \tag{3.22a}$$

and

$$v_1 + v_2 + v_3 + v_4 + v_5 = 2\pi. \tag{3.22b}$$

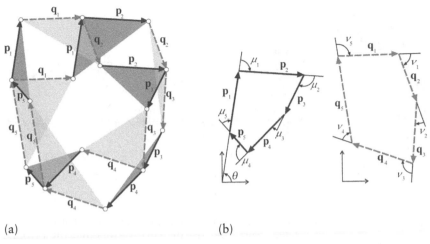

Figure 3.14 (a) A double chain with five pairs: three intersecting and two non-intersecting pairs, and (b) geometrical representation of its mobility conditions.

The conditions for it being closed loop linkage are identical to Eq. (3.10), i.e.

$$\mathbf{p}_1 + \mathbf{q}_2 + \mathbf{p}_2 + \mathbf{q}_3 + \mathbf{p}_3 + \mathbf{q}_4 + \mathbf{p}_4 + \mathbf{q}_5 + \mathbf{p}_5 + \mathbf{q}_1 = \mathbf{0}. \tag{3.23}$$

Using the same approach utilising the complex number notation, we can arrive at the mobility conditions as

$$\mathbf{p}_1 + \mathbf{p}_2 + \mathbf{p}_3 + \mathbf{p}_4 + \mathbf{p}_5 = \mathbf{0}, \tag{3.24a}$$

and

$$\mathbf{q}_1 + \mathbf{q}_2 + \mathbf{q}_3 + \mathbf{q}_4 + \mathbf{q}_5 = \mathbf{0} \tag{3.24b}$$

as indicated in Figure 3.14(b). Hence, the double chain is mobile if Eqs (3.24a) and (3.24b) are satisfied.

Other possible combinations for double chain assembly with five pairs exist. For example, it can have two intersecting pairs followed by three non-intersecting pairs, as shown in Figure 3.15. Let us now investigate the mobility of this closed double chain.

To preserve the parallelogram constraint, the diagrams shown in Figure 13.15(b) can be drawn. There must be

$$\mu_1 + \mu_2 + \mu_3 + \mu_4 + \mu_5 + \theta = 2\pi, \tag{3.25a}$$

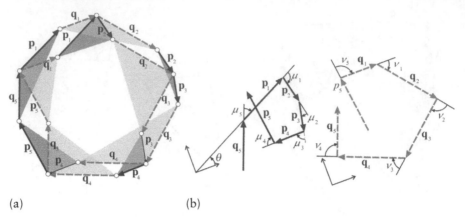

(a)   (b)

Figure 3.15 (a) Double chain with five pairs: three non-intersecting pairs followed by two intersecting pairs, and (b) the edge vectors and the pivoting angle.

and

$$v_1 + v_2 + v_3 + v_4 + v_5 - \theta = 2\pi. \tag{3.25b}$$

Using the same argument for the immobile double chain assembly consisting of five non-intersecting elements, it can be concluded that this assembly is also immobile.

For all of the double chain assemblies made from five pairs, we can determine their mobility using the approach outlined above. The result is given in Table 3.1. Note that for an immobile double chain the total number of non-intersecting pairs is always *odd* regardless of the order of these pairs in the chain.

The above proofs can be extended to double chains with any number of intersecting and non-intersecting elements. To sum up, the mobility conditions of a closed double chain linkage constructed subjected to the loop parallelogram constraint are as follows.

a   the number of non-intersecting pairs must be even; and
b   the sum of vectors **p** and **q** must be zero, respectively.

Based on the derivation, we are able to produce mobile double chains with both an even and odd number of intersecting scissor-like elements. It is particularly interesting to note that we are able to produce a mobile double chain consisting of three intersecting elements, see Figure 3.16.

The limitation of the above derivation is the imposition of the loop parallelogram constraint on all of the double chains. Other mobile closed double chains also exist, which have been briefly outlined in Section 3.1. In

Table 3.1 Five pair double chain assemblies

Cases	Mobility
5 intersecting pairs	Yes
4 intersecting + 1 non-intersecting pairs	No
3 intersecting + 2 non-intersecting pairs	Yes
2 intersecting +1 non-intersecting + 1 intersecting + 1 non-intersecting pairs	Yes
2 intersecting + 3 non-intersecting pairs	No
1 intersecting + 4 non-intersecting pairs	Yes
1 intersecting + 1 non-intersecting + 1 intersecting + 2 non-intersecting pairs	No
5 non-intersecting pairs	No

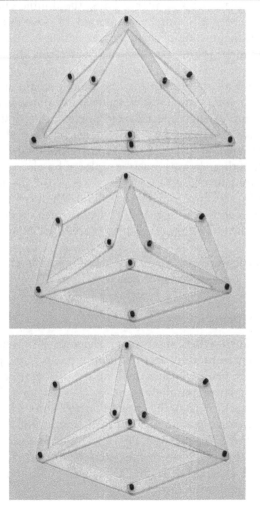

Figure 3.16 Motion sequence of a double chain made of three intersecting pairs.

fact Hoberman (1990, 1991) as well as You and Pellegrino (1997a) have discovered that, by maintaining a constant sustained angle for each scissor-like element or for each set of elements, mobile double chains can be created using symmetry. No loop parallelogram constraint is forced upon the resulting linkages. Figure 3.5(b) shows such an example. Finding a general solution for those cases remains a challenge.

3.3 Supports for double chains

3.3.1 Double chains with a symmetric layout

Double chains with symmetric layout can be connected to supports that permit translation within the plane of symmetry. Thus, the double chain with two-fold symmetry, Figure 3.17, can be supported by four tracks along the symmetry lines. The magnitude of the edge translation is small in comparison with that of the inner joints, because each outer ring distorts less than the inner ring.

The second, less intuitive way of supporting the double chain is to connect its elements to fixed points, which allow rotation but not translation. The existence and location of such special fixed points are easiest to show for regular circular layouts. Figure 3.18 shows such an example in which each of the light grey angulated beams has a corresponding fixed point. To facilitate the rotation these beams must be replaced by plates, e.g. beams $A_1B_1C_1$ and $A_2B_2C_2$ are substituted by larger plates $A_1B_1C_1D_1$ and $A_2B_2C_2D_2$, respectively, and D_1 and D_2 are fixed to ground. The expansion sequence of the double chain is shown in Figure 3.19 from which it can be seen that all of the light grey beams or plates rotate whereas the other set of dark beams translate.

In fact, this is a common feature for all of the mobile double chains obtained in Section 3.2. The proof is given in the next section.

3.3.2 Fixed points for closed double chain

A closed double chain consisting of three intersecting elements is shown in Figure 3.20. The beams are represented by vectors $(\mathbf{p}_1, \mathbf{p}_2)$, $(\mathbf{p}_2, \mathbf{p}_3)$, $(\mathbf{p}_3, \mathbf{p}_1)$, $(\mathbf{q}_1, \mathbf{q}_2)$, $(\mathbf{q}_2, \mathbf{q}_3)$ and $(\mathbf{q}_3, \mathbf{q}_1)$. The chain satisfies the loop parallelogram constraint and therefore there are three parallelograms in the assembly. We have shown in the previous section that the mobility conditions for the assembly are that both sets of vector $\mathbf{p}$s and $\mathbf{q}$s must form a polygon.

Now assume that a set of three *fixed points*, D_i ($i = 1$, 2 and 3), exist for beams $(\mathbf{p}_1, \mathbf{p}_2)$, $(\mathbf{p}_2, \mathbf{p}_3)$ and $(\mathbf{p}_3, \mathbf{p}_1)$, respectively, and $\mathbf{d}_1$, $\mathbf{d}_2$ and $\mathbf{d}_3$ are vectors linking the fixed points D_3 and D_1, D_1 and D_2 as well as D_2 and D_3, respectively. The loop closure equations must be

$$\mathbf{q}_i + \mathbf{p}_i - \overline{D_i A_i} - \mathbf{d}_i + \overline{D_{i-1} A_{i-1}} = 0, \tag{3.26}$$

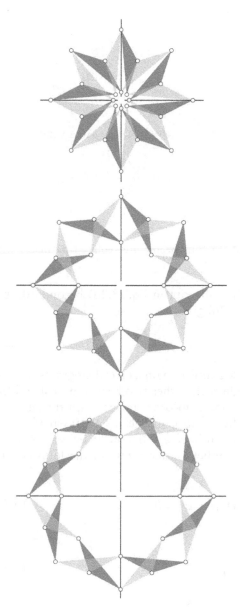

Figure 3.17 Expansion sequence of a double chain with two-fold symmetry.

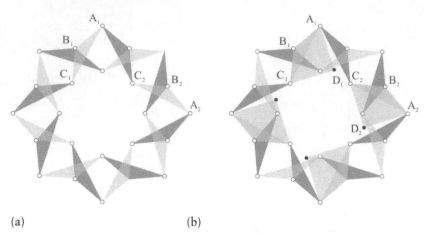

(a)   (b)

Figure 3.18 (a) A double chain and (b) four beams are replaced by plates with fixed points.

in which $i = 1$, 2 and 3. When the subscript in Eq. (3.26) becomes 0, it is replaced by 3. Rearranging Eq. (3.26) gives

$$\mathbf{q}_i - \overline{D_i A_i} + \overline{D_{i-1} A_{i-1}} = \mathbf{d}_i - \mathbf{p}_i. \tag{3.27}$$

During the motion, three parallelograms remain as parallelograms. Hence, if beam $(\mathbf{p}_1, \mathbf{p}_2)$ rotates by an angle ϕ, the other two beams $(\mathbf{p}_2, \mathbf{p}_3)$ and $(\mathbf{p}_3, \mathbf{p}_1)$ have to rotate by the same amount in order to maintain the shape of the parallelograms. Similarly, the rotation of beams represented by qs must be the same, too, though it can have a different value, say ψ. Thus, using the notation of complex numbers, after rotation, the loop closure equation becomes

$$\mathbf{q}_i e^{j\psi} + \mathbf{p}_i e^{j\phi} - \overline{D_i A_i} e^{j\phi} - \mathbf{d}_i + \overline{D_{i-1} A_{i-1}} e^{j\phi} = 0, \tag{3.28}$$

or

$$\left(\mathbf{q}_i - \overline{D_i A_i} + \overline{D_{i-1} A_{i-1}}\right) e^{j\phi} = \mathbf{d}_i - \mathbf{p}_i e^{j\psi}. \tag{3.29}$$

Substituting Eq. (3.27) into Eq. (3.29) yields

$$\mathbf{d}_i (1 - e^{j\phi}) - \mathbf{p}_i (e^{j\psi} - e^{j\phi}) = 0. \tag{3.30}$$

Because $\mathbf{p}_i \neq 0$ and ϕ is completely arbitrary, the possible solutions to Eq. (3.30) are as follows.

$$\mathbf{d}_i = 0 \text{ and } \phi = \psi, \tag{3.31a, b}$$

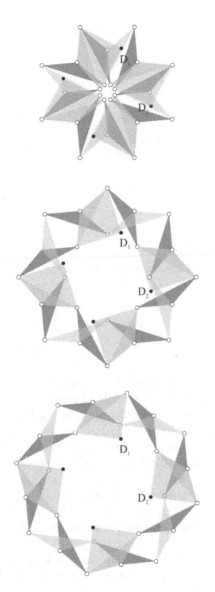

Figure 3.19 Expansion sequence of a double chain with fixed points.

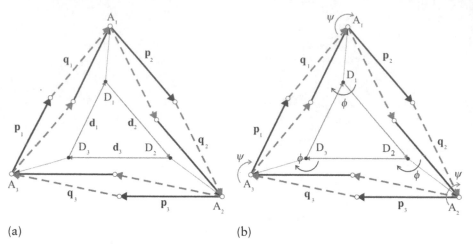

(a)   (b)

Figure 3.20 (a) A double chain with three intersecting elements and (b) the fixing
points and the rotations of beams.

or

$$\mathbf{d}_i = \mathbf{p}_i \text{ and } \psi = 0. \tag{3.32a, b}$$

The first solution indicates that three fixed points have merged into one
and the entire double chain is rotating about a single fixed point. Since we
have not fixed the assembly to a stationary ground this motion is allowed.

The second solution provides the condition for fixed points. Eq. (3.32a)
indicates that the fixed points form a polygon which is identical to the
polygon formed by vectors **ps**. Moreover, the location of the polygon is
not specified and it can be anywhere in the plane. The motion sequence of
a model consisting of three intersecting elements is shown in Figure 3.21.
While half of the beams defined by vectors **ps** rotate about their respective
fixed points during motion, the other half translate without any rotation
because of Eq. (3.32b).

This solution can be extended to closed loop double chain linkages consisting of any number of intersecting elements. Similar solutions can be
found for mobile double chains made of an even number of non-
intersecting elements or a combination of intersecting and non-intersecting
elements. Using the same approach it can also be shown that half of the
rigid beams have fixed points.

3.4 Growth of a double chain

The mobile double chains can be extended by the addition of a pair of bars
of any length, connected to one another and to the double chain by hinges.

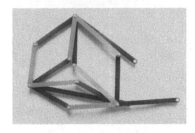

Figure 3.21 Motion sequence of a double chain made of three intersecting pairs. Three of the beams rotate about their respective fixed points on the right.

The resulting structure will remain mobile, like the original double chain, provided that the members added to it are not collinear. Repeating the same arrangement it can be shown that any number of pairs of bars connected by hinges to the double chain will leave its mobility unchanged despite that the chain grows larger. This is best illustrated by the following example.

Figure 3.22(a) shows a general, small part of a double chain consisting of angulated elements. Additional bars are connected to its outer hinges, Figure 3.22(b). It is obvious that the mobility is retained because all additional members are free to rotate with respect to the original double chain. If the lengths of the additional bars are made such that the quadrangles $A_2A_3B_1B_2$, $B_2B_3C_2C_1$, etc. are parallelograms, $\angle A_1A_2A_3$, $\angle B_1B_2B_3$, etc. remain constant when the double chain is in motion. Consider, for example, $\angle A_1A_2A_3$. Because A_1A_2 and A_2A_3 remain parallel to B_0B_1 and B_1B_2, respectively,

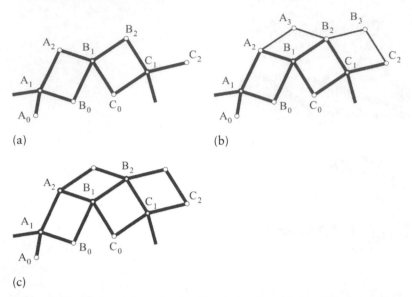

(a)   (b)

(c)

Figure 3.22 (a) A portion of a double chain, (b) additional bars are added and (c) formation of multiple angulated beams.

$$\angle A_1A_2A_3 = \angle B_0B_1B_2 = \text{constant} \tag{3.33}$$

since $\angle B_0B_1B_2$ is the kink angle of an angulated piece, which is fixed. In other words, no relative rotation between angulated beam $A_0A_1A_2$ and the added bar A_2A_3 occurs as the chain moves. The added bar can therefore be bonded to the angulated beam forming a longer beam $A_0A_1A_2A_3$ with two kinks. The same applies to other beams and their respective added bars.

   In conclusion, following the procedure given above the double closed chain can extended to a larger assembly. Kinked beams with single kink can be replaced by beams with more kinks and there is a hinge at every kink. The double chain grows to a multiple chain. Figure 3.23 shows two symmetrical mobile chains whose internal hinges form elliptical and rectangular shapes, respectively. Note that portions of the second chain are made of conventional scissor-like elements.

3.5 Conclusions

The construction of mobile double chains and their extension, the multiple chains, have great potential for applications as retractable roofs. The ample freedom in design allows one to engineer assembly that fits the required space with adequate supports. When a chain is fully expanded, all of the beams move towards its perimeter, and when fully contracted it can

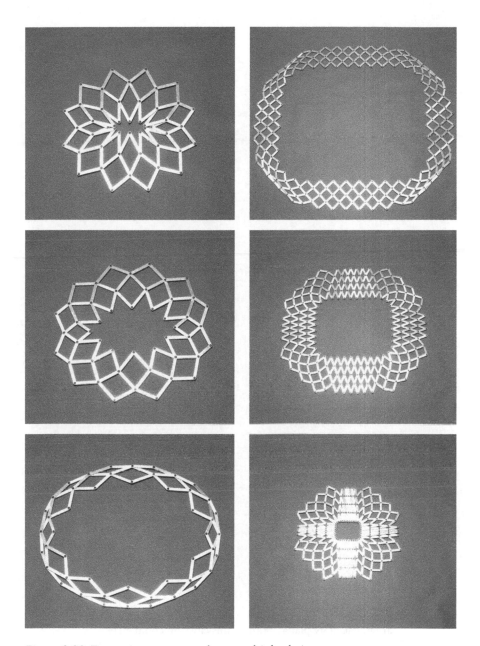

Figure 3.23 Expansion sequence of two multiple chains.

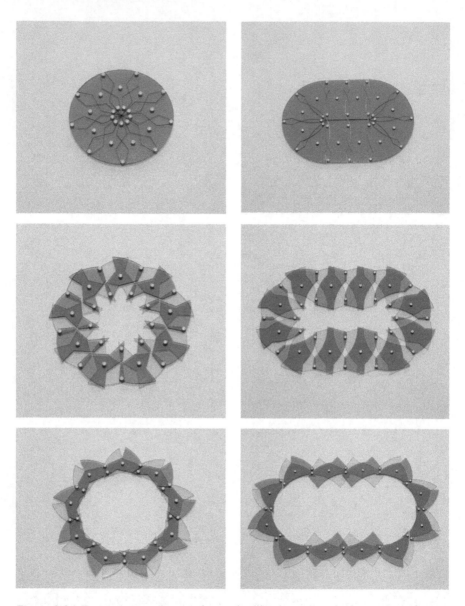

Figure 3.24 Expansion sequence of two double chains in which angulated beams are replaced by plates. Both chains close up completely.

span the entire inner space. In plan, these chains are formed by a series of continuous beams with multiple kinks, which, in practice, can be made by other rigid structures with span and depth to provide sufficient stiffness and coverage as long as all the hinges are at the right places. Kassabian *et al.* (1999), Jensen and Pellegrino (2002) and Luo *et al.* (2007) proposed methods for replacing the angulated beams with rigid plates that can provide complete coverage. Readers may refer to their articles for details. Here we show two of the assemblies devised by their approach in Figure 3.24 where all the plates neatly meet with one another once the assemblies fold. Various other ways of designing plates also exist if overlapping of plates is allowed. This is only practical provided that the adjacent plates are placed at different heights to avoid physical collision of the plates during motion.

4 Spatial rings and domes

DOI: 10.1201/9781482266610-4

4.1 Introduction

The conventional scissor-like element is one of the most popular mechanisms utilised to form large foldable assemblies like pop-up display stands, folding chairs, portable shelters and even swimming pool covers. However, readers should be aware that some of the assemblies are in fact deformable structures, as the expansion induces strain within structural components. They further differ from normal deformable structures in that the strain in these structures falls to zero in the initial folded and final expanded configurations, resulting in self-locking once being completely folded or expanded because any shape change away from these zero-strain configurations requires energy. It is difficult to find a generic design solution for this type of assemblies since detailed analysis at every expansion configuration is required to ensure that the stress does not exceed appropriate limits. Examples of this group of assemblies include Zeigler's dome (1981), structures proposed by Gantes (1991) and a swimming pool cover by Escrig *et al.* (1996).

The number of strain-free motion assemblies using the conventional scissor-like element is rather limited. Typically they are open chains, with the notable exception of the domes proposed by Meurant (1993) and Sánchez-Cuenca (1996). Since we have shown in Section 3.1 that the conventional scissor-like element alone cannot be used to build a mobile planar closed chain which has mobility, the mobile assembly always takes a spatial form. This only becomes possible with the utilisation of the Sarrus linkage for connection of planar chains of conventional scissor-like elements.

The Sarrus linkage (Sarrus, 1853) was the first reported spatial overconstrained linkage. It consists of six links, Figure 2.11 and reproduced in Figure 4.1, connected together by two sets of three parallel hinges. A simple spatial closed chain of four conventional scissor-like elements can be constructed as shown in Figure 4.2(a). These elements are identical and have pivots at the middle of the straight rods. At each of the corners of the assembly, the upper and lower joint pieces and four beams that are hinged together by them form a Sarrus linkage, which are given by their axes 1,

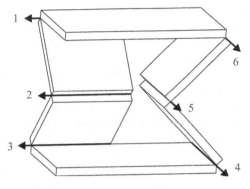

Figure 4.1 A Sarrus linkage where its axes, 1, 2, ..., 6, are represented by arrows.

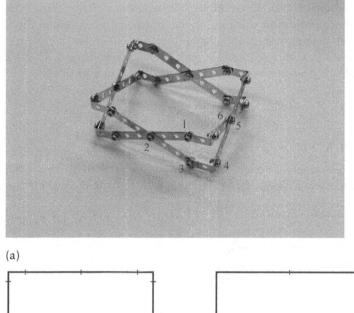

(a)

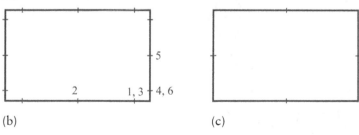

(b)   (c)

Figure 4.2 (a) A spatial closed chain consisting of four conventional scissor-like elements. There is a Sarrus linkage whose axes are marked as 1, 2, ..., 6 at each corner; (b) schematic diagram representing the top projection of the spatial chain; and (c) the same diagram when joint size is neglected.

2,..., 6 in Figure 4.2(a). All of the connections are revolute joints and the chain folds to a compact bundle.

The chain has a total of sixteen links ($n = 16$), including eight joint pieces, and twenty revolute joints ($j = 20$), each of which has one degree of freedom ($f_i = 1$). The mobility criterion given by Eq. (2.1) therefore yields a value of -10. However, as all of the scissor-like elements can have the same pivoting angles because of symmetry it is an overconstrained mechanism. Its top projection on the ground is always a rectangle as shown by the schematic diagram in Figure 4.2(b) in which the thick solid lines are projections of elements and little dashes represent revolute joints. Moreover, if the dimension of the joints is considerably smaller than that of the elements, the schematic diagram becomes that in Figure 4.2(c). Note that, though the joint dimensions may be neglected as far as the overall assembly is concerned, they have to be taken into account at a later stage in order to ensure the assembly is a true mechanism.

The above method for construction of the spatial mobile chain can be extended to include more or fewer conventional scissor-like elements. For instance, a closed chain with three scissor-like elements of the same type can be made whose projection during deployment becomes equilateral triangles. Using this assembly as a unit, a mobile ring assembly whose top projection is shown in Figure 4.3 can be built (You and Pellegrino, 1997b). The variation of this design is the basis for most of the pop-up stands.

4.2 Rings

4.2.1 *Formation of rings*

The conventional scissor-like elements can also be used to form spatial rings. The process is as follows. First, we build a closed chain consisting of

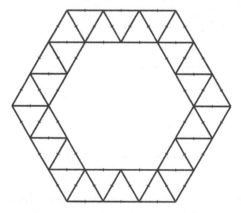

Figure 4.3 Projection of a large assembly made of conventional scissor-like elements. The short dashes represent mid pivots of the elements.

n_a straight scissor-like elements that are made of rods of equal length $2a$ with a pivot in the middle. A similar closed chain can be formed by n_b elements with rod length of $2b$ with again a middle pivot. Both chains are symmetric about the middle plane on which all middle pivots of scissor-like elements lie. Obviously, both of the chains could become very flexible while n_a and n_b are large because of the tolerance in each revolute. One way to stiffen the structure is to put one chain inside of the other and then connect them with a number of *intermediate ties* which are also conventional scissor-like elements to form an integrated larger assembly. The question that remains to be answered is what the conditions are for the assembly to retain mobility one.

Name the first chain as the inner loop and the second the outer loop. For the inner loop, each of the scissor-like elements occupies a sector with a corresponding central angle

$$\alpha = \frac{2\pi}{n_a}.\tag{4.1}$$

Denote by θ the pivoting angle for elements in the inner loop and let the elements in the outer loop have the *same* deployment angle, see Figure 4.4(a) and (b). The projection lengths of the inner and outer elements, L_a and L_b, are

$$L_a = 2a\sin\frac{\theta}{2} \text{ and } L_b = 2b\sin\frac{\theta}{2},\tag{4.2a, b}$$

respectively, whereas the heights of the elements, H_a and H_b, are

$$H_a = 2a\cos\frac{\theta}{2} \text{ and } H_b = 2b\cos\frac{\theta}{2}.\tag{4.3a, b}$$

To effectively connect the two loops with evenly distributed sets of elements, n_b should be either n_a or $2n_a$. Three different ways of arranging these elements have been found. Figure 4.5 shows the top projection view

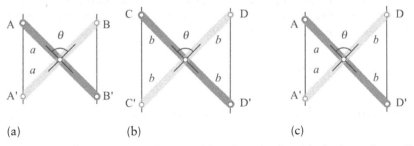

(a)   (b)   (c)

Figure 4.4 The conventional scissor-like elements for (a) the inner loop, (b) the outer loop and (c) the intermediate tie.

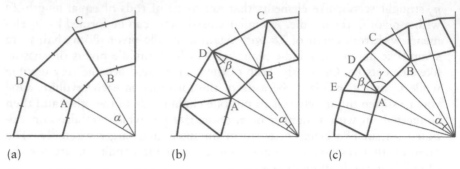

| (a) | (b) | (c) |

Figure 4.5 Projections of (a) ring concept A, (b) ring concept B and (c) ring concept C. Only a quarter of the rings are shown.

of three rings, referred to as concepts A, B and C hereafter. The first two have $n_b = n_a$ and their top projections consist of n_a trapezia and $2n_a$ isosceles triangles, respectively, but for the third one, $n_b = 2n_a$, and the projection has n_a trapezia and n_a isosceles triangles.

To ensure the mobility of the rings, further geometrical conditions have to be met, which are discussed in next section.

4.2.2 Concept A

This is the simplest of the three rings in which the inner and outer loops have the same number of elements and they are connected by a set of n_a intermediate ties forming a ring of n_a trapezia. The projection of a quarter of the ring is shown in Figure 4.5(a). In a single trapezium ABCD, AB and CD are a pair of beams of lengths $2a$ and $2b$, respectively. BC and AD are the projection of the intermediate tie which is also a conventional scissor-like element. This element consists of a pair of rods $a+b$, Figure 4.4(c), with the semi-length a near the inner loop and semi-length b near the outer loop so that the heights of the element are equal to H_a and H_b given in Eqs (4.3a) and (4.3b), respectively. This allows a connection between the inner and outer loops providing that the pivoting angle of the intermediate element is also θ. To put together a ring of n_a identical modules there is a further geometric condition, to ensure each module remains within a sector with a subtended central angle α. Therefore,

$$2AD \sin \frac{\alpha}{2} = CD - AB. \tag{4.4}$$

Expressing AB, CD and AD in terms of the beam lengths and pivoting angle,

$$AB = 2a \sin \frac{\theta}{2}, \quad CD = 2b \sin \frac{\theta}{2} \quad \text{and} \quad AD = (a+b) \sin \frac{\theta}{2}. \tag{4.5a, b, c}$$

Substituting the above equations into Eq. (4.4) yields

$$2(a+b)\sin\frac{\theta}{2}\sin\frac{\alpha}{2} = 2b\sin\frac{\theta}{2} - 2a\sin\frac{\theta}{2},$$

from which

$$\frac{b}{a} = \frac{1+\sin\dfrac{\alpha}{2}}{1-\sin\dfrac{\alpha}{2}}. \tag{4.6}$$

Ratio b/a is a constant for a given n_a due to Eq. (4.1).

Eq. (4.6) indicates that in order for the inner loop and outer loop to be connected with a single intermediate element shown in Figure 4.4(c), the ratio between the lengths of the elements forming the inner and outer loops must be kept constant once the number of elements and the layout of ring are determined.

4.2.3 Concept B

The project of this ring consists of $2n_a$ triangles. The plan view, Figure 4.5(b), is a chain of isosceles triangles with alternate long bases, e.g. CD, corresponding to a pair of rods of length $2b$, and short bases, e.g. AB, corresponding to a pair of rods of length $2a$. All the intermediate scissor-like elements connecting the inner and outer loops are of length $a+b$ with the semi-length a near the inner loop just as those for concept A. The heights of the intermediate elements match those of the elements in the inner and outer loop, hence all that remains to be done is to find the geometric condition to complete the assembly. Here

$$AB = 2AD\sin\frac{\beta}{2} \tag{4.7}$$

and

$$CD = 2AD\sin\frac{\alpha+\beta}{2}. \tag{4.8}$$

During motion,

$$AB = 2a\sin\frac{\theta}{2}, \quad CD = 2b\sin\frac{\theta}{2} \quad \text{and} \quad AD = (a+b)\sin\frac{\theta}{2}. \tag{4.9a, b, c}$$

Substituting these expressions into Eqs (4.7) and (4.8) gives two equations in terms of a, b and β, from which

$$\frac{b}{a} = \frac{\cos\dfrac{\alpha}{2} + \sin\dfrac{\alpha}{2}\left(\sin\dfrac{\alpha}{2} + \sqrt{1 + 2\cos\dfrac{\alpha}{2}}\right)}{\cos^2\dfrac{\alpha}{2}} \tag{4.10}$$

and

$$\beta = 2\arcsin\frac{a}{a+b}. \tag{4.11}$$

4.2.4 Concept C

This ring is made from n_a trapezia and n_a triangles arranged so that $2n_a$ scissor-like elements with beam length $2b$ form the outside of the ring, while only n_a scissor-like elements with rod length $2a$ form the inner loop. In plan view, Figure 4.5(c), this results in a chain of alternate isosceles triangles and isosceles trapezia. The intermediate elements are the same type as those used in concepts A and B whose heights meet those of the elements to be connected.

It can be shown that

$$\frac{1}{2}DE = AD\sin\frac{\beta}{2} \tag{4.12}$$

and

$$AB = CD + 2AD\cos\gamma, \tag{4.13}$$

where $\gamma = \pi - \dfrac{\beta}{2} - \left(\dfrac{\pi}{2} - \dfrac{\alpha}{2}\right) = \dfrac{\pi}{2} - \dfrac{\beta - \alpha}{2}$. Noting that

$$AB = 2a\sin\frac{\theta}{2}, \quad CD = 2b\sin\frac{\theta}{2} \quad \text{and} \quad AD = (a+b)\sin\frac{\theta}{2}, \tag{4.14a, b, c}$$

we have

$$\frac{b}{a} = \frac{\left(1 + \sqrt{2}\sin\dfrac{\alpha}{4}\right)^2}{2\cos^2\dfrac{\alpha}{4}} \tag{4.15}$$

and

$$\beta = 2\arcsin\frac{b}{a+b}.$$ (4.16)

4.2.5 Comparison

The three concepts for the mobile rings made from the conventional scissor-like elements lead to a family of rings[1] of mobility one. All of them can be folded to a compact bundle when pivoting angle $\theta = 0$ making them ideal for applications where compact packaging is essential. It is also important to note that a and b, which define the lengths of the beams, cannot be independently chosen as the ratio b/a is dependent upon on the layout of the ring and n_a.

Selection of the rings depends on the actual applications. For example, when the ring is for a support structure of a deployable reflector, the beams are likely to carry compressive forces. The most effective restraint against buckling is achieved for $b/a \approx 1$.

Figure 4.6, obtained by plotting Eqs (4.6), (4.10) and (4.15) in which α is replaced by n_a from Eq. (4.1), shows that concepts A and B are almost equivalent in this respect, but concept C is much better for smaller n_a.

The model of Figure 4.7 was made for $n_a = 12$ and it was based on concept B with $b/a = 1.582$. Concept C gives $b/a = 0.714$, in which a and b seem to be closer. However, it requires far more joints than the other two concepts.

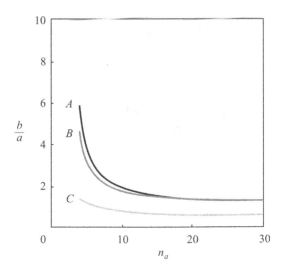

Figure 4.6 b/a vs n_a.

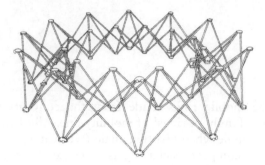

Figure 4.7 A ring based on concept B with $n_a = 12$.

4.3 Domes

4.3.1 *Generalised rings*

A ring structure can be extended by adding more outer loops connected by intermediate elements. The intermediate elements between any two neighbouring loops are decided using one of the concepts presented in the previous section. For instance, a ring based on concept B can be put outside of a ring of concept C to form a mobile assembly shown in Figure 4.8. The

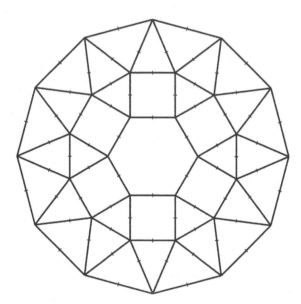

Figure 4.8 Projection of a generalised ring obtained by having a concept B ring wrapped around by a concept C ring.

profile of assembly is determined by the ratios given in Eqs (4.10) and (4.16). It is in general unable to form a curved dome-like profile when expanded because the newly created assembly is symmetric about its middle plane on which the middle pivots of all the scissor-like elements lie just like the individual rings. To form a proper dome, the outer loop must differ in height from the inner loop, i.e. the loops need to become higher and higher from the outer perimeter towards the centre. This can be achieved by using an alterative type of intermediate tie.

4.3.2 Alternative intermediate tie

The alternative intermediate tie, Figure 4.9(a), is made of a chain of two conventional scissor-like elements where HM and OJ are parallel to each other and so are NJ and MI. The tie has a single mobility defined by pivoting angle θ: $\angle ILO = \angle HKN = \theta$. Moreover,

$$HM = MI = a \text{ and } NJ = OJ = b \tag{4.17}$$

so that both HI and NO are parallel, and the heights of the tie are equal to H_a and H_b given in Eqs (4.3a) and (4.3b), respectively, allowing it to be connected to the elements in inner and outer loops.

Let $MK = KJ = c$. From Figure 4.9(a) the projection of the new element is

$$\left(NJ - KJ\right)\sin\frac{\theta}{2} + \left(HM - MK\right)\sin\frac{\theta}{2} = \left(a + b - 2c\right)\sin\frac{\theta}{2}. \tag{4.18}$$

For the layouts of concepts A, B and C, to replace the single element ties with the alternative one, the above projection length must be equal to AD in Figure 4.5, i.e.

$$AD = \left(a + b - 2c\right)\sin\frac{\theta}{2}. \tag{4.19}$$

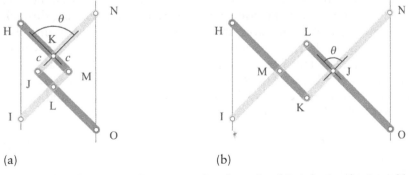

(a)   (b)

Figure 4.9 (a) A intermediate tie made of a pair of conventional scissor-like elements and (b) its variation.

Replacing AD in Eqs (4.5c), (4.9c) and (4.14c) with the one given in Eq. (4.19) and then following the same subsequent derivations, we obtain

$$c = \frac{a+b-l}{2} \tag{4.20}$$

where

$$l = \frac{b-a}{\sin\dfrac{\alpha}{2}}, \tag{4.21}$$

$$l = \frac{\sqrt{a^2 + b^2 - 2ab\cos\dfrac{\alpha}{2}}}{\sin\dfrac{\alpha}{2}}, \tag{4.22}$$

and

$$l = \frac{\sqrt{a^2 + 4b^2\cos^2\dfrac{\alpha}{4} - 4ab\cos^2\dfrac{\alpha}{4}}}{\sin\dfrac{\alpha}{2}} \tag{4.23}$$

for layouts of concepts A, B and C, respectively.

Obtained from Eq. (4.20), c could be negative in which case the alternative tie has the form shown in Figure 4.9(b). However, c could also be zero in which case the alternative tie reduces to a single scissor-like element. Substituting Eqs (4.21), (4.22) or (4.23) into Eq. (4.20) and letting $c=0$, ratio b/a can be obtained which is identical to the ones given by Eqs (4.6), (4.10) and (4.15).

The above alternative tie allows both a and b being selected independently. It has yet to enable us to alter the heights of the end connectors because the tie remains symmetric about a horizontal line passing through J and M, see Figure 4.9. However, this can be done by shifting beams HM and MI in Figure 4.9(a) vertically, resulting in the elements shown in Figure 4.10(a). Note that MK and KJ become different in length after this action, though $\angle HKN = \angle ILO = \theta$, both HI and NO are still parallel and the heights of the tie are equal to H_a and H_b given in Eqs (4.3a) and (4.3b). Denote $MK = c'$ and $KJ = c''$. Eq. (4.20) becomes

$$c' + c'' = a + b - l. \tag{4.24}$$

The shifting h, the difference of height of current and previous position of node H, or I, is

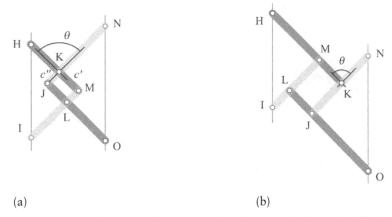

(a)　　　　　　　　　　　　　　(b)

Figure 4.10 (a) A intermediate tie made of a pair of conventional scissor-like elements with height shift and (b) its variation.

$$h = (c' - c'')\cos\frac{\theta}{2}. \tag{4.25}$$

Therefore, from Eqs (4.24) and (4.25), c' and c'' can be expressed in terms of h and l,

$$c', c'' = 0.5\left(a + b - l \mp \frac{h}{\cos\dfrac{\theta}{2}}\right). \tag{4.26}$$

To achieve a height differentiation h at a particular expanded configuration when $\theta = \theta_f$, c' and c'' can be determined from Eq. (4.26). Again they can be negative in which case the tie appears similar to that of Figure 4.10(b).

4.3.3 Construction of domes

The alternative intermediate tie gives greater design flexibility. One can either select a layout among those in Figure 4.5 for a ring or combine them for a dome. The primary design variables for rings are n_a, a and b. The domes require two additional design variables: h, the height differentiation between any of two neighbouring loops, and θ_f, corresponded to the expanded configuration at which h is achieved. All of the dimensions of the assembly can be obtained accordingly once those design variables are known.

Figure 4.11 shows the expansion sequence of a mobile frame for a dome obtained by combining a concept B ring with one based on concept C using a layout identical to that shown in Figure 4.8. From a bundle it expands to a dome shape. It can also be folded to flat, revealing clearly its layout.

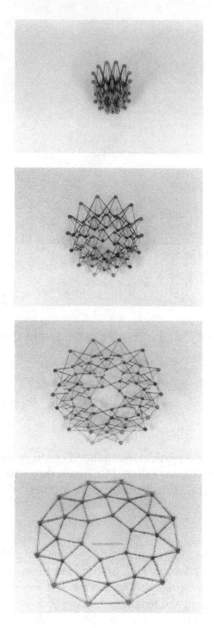

Figure 4.11 Expansion sequence of a mobile dome frame.

4.4 Other design considerations

4.4.1 Corner connectors

At the corners of either a ring or a dome, corner joint pieces are used where beams from the scissor-like elements in a loop and tie meet. The upper and lower pieces, together with beams being connected, form a Sarrus linkage. The derivations for both rings and domes are based on an implicit assumption that the links and the joints have no physical size. This is true if the axes of all revolute joints meet at a single point but this is not always possible.

The derivations in the previous sections can be modified to include the distances between the axes of end connectors of scissor-like elements and the point where all the axes are supposed to meet, known as the *eccentricities*. Take a ring based on concept A as an example. Let Λ and Δ be the eccentricities of corner joints pieces at outer loop and λ and Δ be those of pieces at inner loop, see Figure 4.12. The projection lengths given in Eq. (4.5) need to be modified to reflect them. Now,

$$AB = 2a\sin\frac{\theta}{2}+2\lambda, \quad CD = 2b\sin\frac{\theta}{2}+2\Lambda \quad \text{and} \quad AD = (a+b)\sin\frac{\theta}{2}+\delta+\Delta.$$

$$(4.27\text{a, b, c})$$

Substituting the above expressions into Eq. (4.4) gives

$$2(a+b)\sin\frac{\theta}{2}\sin\frac{\alpha}{2}+2(\delta+\Delta)\sin\frac{\alpha}{2} = 2b\sin\frac{\theta}{2}-2a\sin\frac{\theta}{2}+2(\Lambda-\lambda), \quad (4.28)$$

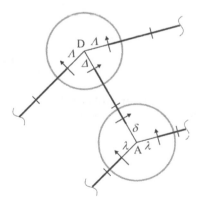

Figure 4.12 Projection of a portion of a ring based on concept A with eccentricities at the corner joint pieces. Arrows represent the axes of the revolute joints at the ends of scissor-like elements.

which can be broken down into

$$\left[(a+b)\sin\frac{\alpha}{2}-(b-a)\right]\sin\frac{\theta}{2}=0 \tag{4.29}$$

and

$$(\delta+\Delta)\sin\frac{\alpha}{2}-(\Lambda-\lambda)=0. \tag{4.30}$$

Eq. (4.29) is identical to Eq. (4.6) whereas Eq. (4.30) is the equation that determines the relationship among eccentricities. One possible solution to Eq. (4.30) is

$$\lambda:\Lambda:(\delta+\Delta)=a:b:(a+b) \tag{4.31}$$

i.e. the eccentricities are proportional to the length of the rods.

4.4.2 Diagonal ties

Rings based on concepts A and C are in general less stiff than those using concept B owing to the trapezia in the layout instead of triangles. However, it is possible to triangulate a trapezium by having an alternative intermediate tie diagonally placed. In concept A layout shown in Figure 4.5(a), for instance, A and C could be bridged by such a tie. The projection length of AC is

$$AC=l\sin\frac{\theta}{2}, \tag{4.32}$$

in which

$$l=\sqrt{4a^2+(a+b)^2+4a(a+b)\sin\frac{\alpha}{2}}. \tag{4.33}$$

The alternative intermediate tie can be designed by substituting l into Eq. (4.20). It forms an integrated part of the ring just like any other elements. The entire assembly remains mobile.

5 Spatial motion structures based on the Bennett linkage

DOI: 10.1201/9781482266610-5

5.1 Introduction

Designing a large motion structure has different priorities from designing a machine. There are two keys to a successful concept. First, to identify one of a small number of robust and scalable building blocks made of known mechanisms, and, second, to develop a way by which the building blocks can be connected to form a large assembly while retaining the mobility of each mechanism.

Motion structures developed over the past three decades are based mostly on planar mechanisms, such as those introduced in Chapters 3 and 4. It is rare to use truly three dimensional mechanisms. The reason is primarily because the majority of structural engineers are less familiar with the three dimensional mechanisms. In this chapter, we shall introduce types of motion structures that are constructed by tessellation of the Bennett linkage.

The kinematics of Bennett linkage has been summarised in Chapter 2, including its geometrical features and closure equations. As a spatial four-bar linkage, a typical Bennett linkage, Figure 5.1(a), consists of a closed chain of four bars which span the shortest distance between two axes of adjacent revolute joints. Each of the bars has lengths and twists identical to those of the bar which is not directly connected to it. It therefore can be represented schematically by a rectangle shown in Figure 5.1(b). Each side of the rectangle corresponds to a link; at every corner where two links meet is a revolute joint represented by a black dot. Twists of each link are given alongside the respective bars. This simplified diagram will be used throughout this chapter to avoid confusion in drawing real three dimensional assemblies. In the next section we shall use the Bennett linkage as the building block to construct large motion structures.

5.2 Single-layer assembly of Bennett linkages

5.2.1 The layout

A layout of the motion structure is given in Figure 5.2(a) in which a number of Bennett linkages represented by large rectangles similar to that

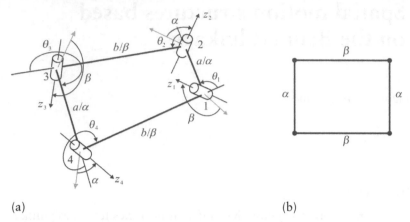

(a)   (b)

Figure 5.1 (a) A Bennett linkage and (b) its schematic diagram.

shown in Figure 5.1 are connected together to form an three dimensional assembly (Chen and You, 2005). Each link is represented by a straight line and there is a revolute joint at every cross between two links to create a total of four revolute joints on each link. The network has many rectangles; each is a 4R closed chain. So hereafter we also refer to them as 4R loops. In order to retain mobility, the 4R loops must be Bennett linkages or otherwise the assembly would be locked.

An enlarged portion of the layout given in Figure 5.2(a) is shown in Figure 5.2(b). Assume that the large rectangle ABCD has link lengths a, b and twists α, β. The smaller rectangles around it, numbered as Bennett linkages 1 to 8, will have lengths will have lengths a_i, b_i and twists α_i, β_i ($i = 1, 2, \ldots, 8$), respectively.

Since AB, BC, CD and DA are single links, there must be

$$a_1 + a_2 + a_3 = a \,,\ b_3 + b_4 + b_5 = b,$$

$$a_5 + a_6 + a_7 = a \,,\ b_7 + b_8 + b_1 = b, \tag{5.1}$$

$$\alpha_1 + \alpha_2 + \alpha_3 = \alpha \,,\ \beta_3 + \beta_4 + \beta_5 = \beta,$$

$$\alpha_5 + \alpha_6 + \alpha_7 = \alpha \,,\ \beta_7 + \beta_8 + \beta_1 = \beta. \tag{5.2}$$

Denote by σ, τ, υ and ϕ the revolute variables. For Bennett linkage 1, we have

$$\tan\frac{\pi - \upsilon}{2}\tan\frac{\pi - \sigma}{2} = \frac{\sin\frac{1}{2}(\beta_1 + \alpha_1)}{\sin\frac{1}{2}(\beta_1 - \alpha_1)}, \tag{5.3}$$

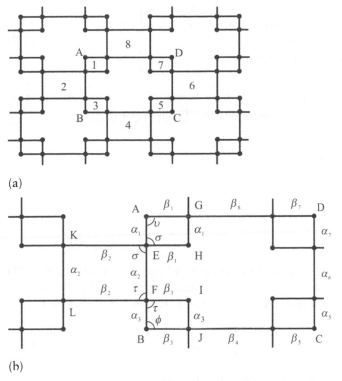

(a)

(b)

Figure 5.2 Single-layer assembly of Bennett linkages. (a) A portion of the network; (b) enlarged connection details.

because of Eq. (2.27c). Similarly, for Bennett linkages 2, 3 and ABCD,

$$\tan\frac{\pi-\sigma}{2}\tan\frac{\pi-\tau}{2}=\frac{\sin\frac{1}{2}(\beta_2+\alpha_2)}{\sin\frac{1}{2}(\beta_2-\alpha_2)}, \qquad (5.4)$$

$$\tan\frac{\pi-\tau}{2}\tan\frac{\pi-\phi}{2}=\frac{\sin\frac{1}{2}(\beta_3+\alpha_3)}{\sin\frac{1}{2}(\beta_3-\alpha_3)}, \qquad (5.5)$$

$$\tan\frac{\pi-\upsilon}{2}\tan\frac{\pi-\phi}{2}=\frac{\sin\frac{1}{2}(\beta+\alpha)}{\sin\frac{1}{2}(\beta-\alpha)}. \qquad (5.6)$$

Combining Eqs (5.3) to (5.6) gives

$$\frac{\sin\frac{1}{2}(\beta_1+\alpha_1)}{\sin\frac{1}{2}(\beta_1-\alpha_1)}\cdot\frac{\sin\frac{1}{2}(\beta_3+\alpha_3)}{\sin\frac{1}{2}(\beta_3-\alpha_3)}=\frac{\sin\frac{1}{2}(\beta+\alpha)}{\sin\frac{1}{2}(\beta-\alpha)}\cdot\frac{\sin\frac{1}{2}(\beta_2+\alpha_2)}{\sin\frac{1}{2}(\beta_2-\alpha_2)}. \tag{5.7}$$

This is a non-linear equation and many solutions may exist. By observation, two solutions can be immediately determined, which are

$$\begin{aligned}\alpha_3&=\alpha, & \alpha_2&=-\alpha_1,\\ \beta_3&=\beta, & \beta_2&=-\beta_1,\end{aligned} \tag{5.8}$$

and

$$\begin{aligned}\alpha_1&=\alpha, & \alpha_2&=-\alpha_3,\\ \beta_1&=\beta, & \beta_2&=-\beta_3.\end{aligned} \tag{5.9}$$

Similar analysis can be applied to Bennett linkages around links BC, CD and DA, see in Figure 5.2(a). Based on solutions (5.8) and (5.9), we can conclude that twists of Bennett linkages 3, 4 and 5 should therefore satisfy

$$\begin{aligned}\alpha_3&=\alpha, & \alpha_4&=-\alpha_5, && \alpha_5=\alpha, & \alpha_4&=-\alpha_3,\\ \beta_3&=\beta, & \beta_4&=-\beta_5, & \text{or}\quad & \beta_5=\beta, & \beta_4&=-\beta_3,\end{aligned} \tag{5.10}$$

twists of Bennett linkages 5, 6 and 7 should satisfy

$$\begin{aligned}\alpha_7&=\alpha, & \alpha_6&=-\alpha_5, && \alpha_5=\alpha, & \alpha_6&=-\alpha_7,\\ \beta_7&=\beta, & \beta_6&=-\beta_5, & \text{or}\quad & \beta_5=\beta, & \beta_6&=-\beta_7,\end{aligned} \tag{5.11}$$

and twists of Bennett linkages 7, 8 and 1 should satisfy

$$\begin{aligned}\alpha_7&=\alpha, & \alpha_8&=-\alpha_1, && \alpha_1=\alpha, & \alpha_8&=-\alpha_7,\\ \beta_7&=\beta, & \beta_8&=-\beta_1, & \text{or}\quad & \beta_1=\beta, & \beta_8&=-\beta_7.\end{aligned} \tag{5.12}$$

Combining four sets of solutions Eqs (5.8) to (5.12), two common solutions which enable the network in Figure 5.2(a) to become mobile, are obtained:

$$\begin{aligned}\alpha_1&=\alpha_5=\alpha, & \alpha_2&=\alpha_4=-\alpha_3, & \alpha_6&=\alpha_8=-\alpha_7,\\ \beta_1&=\beta_5=\beta, & \beta_2&=\beta_4=-\beta_3, & \beta_6&=\beta_8=-\beta_7,\end{aligned} \tag{5.13}$$

and

$$\begin{aligned}\alpha_3&=\alpha_7=\alpha, & \alpha_2&=\alpha_8=-\alpha_1, & \alpha_6&=\alpha_4=-\alpha_5,\\ \beta_3&=\beta_7=\beta, & \beta_2&=\beta_8=-\beta_1, & \beta_6&=\beta_4=-\beta_5.\end{aligned} \tag{5.14}$$

Due to symmetry, Eqs (5.13) and (5.14) are essentially the same. Thus, only Eq. (5.13) is used in the derivation next.

So far, only twists of each loop have been considered. However, due to Eq. (2.24) the corresponding lengths of each Bennett linkage must satisfy

$$\frac{\sin \alpha}{\sin \beta} = \frac{\sin \alpha_i}{\sin \beta_i} = \frac{a}{b} = \frac{a_i}{b_i} \quad (i = 1, 2, \ldots, 8). \tag{5.15}$$

A close examination of Eqs (5.13) and (5.15) reveals the order of distribution of Bennett linkages within the assembly. Diagonally from top left corner to bottom right, the rows of Bennett linkages can be grouped into two types, Figure 5.3(a). The first type consist of Bennett linkages with

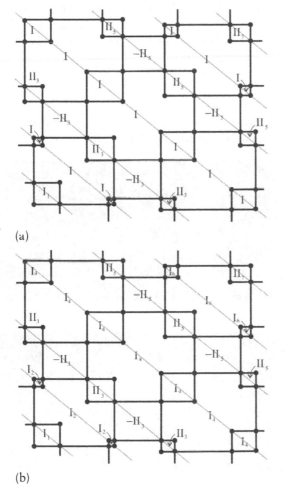

(a)

(b)

Figure 5.3 Mobile assembly of Bennett linkages. (a) A simple case and (b) a more general case. The grey lines are the guidelines along which the Bennett linkages are of the same type.

lengths proportional to a and b, and twists being α and β, which are denoted as 'I' in Figure 5.3(a). The second type, next to the first one, are made of Bennett linkages with lengths proportional to a and b, and twists being α_i and β_i or $-\alpha_i$ and $-\beta_i$, which are denoted as 'II$_i$' or '$-$II$_i$' ($i = 1, 3, 5, \ldots$), respectively, and both α_i and β_i satisfy Eq. (5.15).

A further analysis has shown that the rows of Bennett linkages in Figure 5.3(a) marked 'I' can actually have different twists from other rows where only one type of Bennett linkages is allowed. A more general solution is given in Figure 5.3(b) in which the Bennett linkages 'I' are replaced by 'I$_k$' ($k = 2, 4, 6, \ldots$). A very interesting pattern now emerges: diagonally from top left to bottom right, each row of the Bennett linkages belong to the same type. This leads to the following important feature. If a set of straight and parallel guidelines are drawn diagonally from top left to bottom right, each of which passes through a number of revolute joints, these lines remain straight and parallel to each other during deployment though the distances between the guidelines may vary.

5.2.2 Expanded shapes

In general, the assembly shown in Figure 5.3(b) deploys into a cylindrical profile. Throughout deployment, the guidelines linking the respective revo-

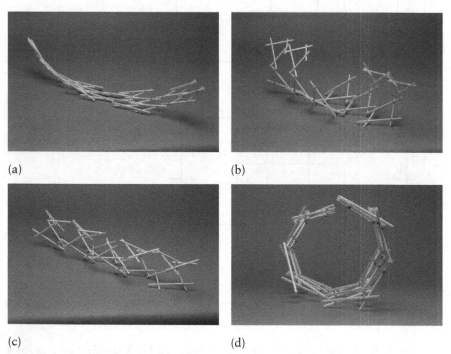

(a)

(b)

(c)

(d)

Figure 5.4 An example of single-layer network of Bennett linkages. (a) to (c) Deployment sequence; (d) view of cross-section of network.

lute joints remain straight. They expand along the longitudinal direction of the cylinder. In the other diagonal direction, i.e. the direction from top right to bottom left, joints generally deploy *spirally* on the surface of the cylinder.

A model is shown in Figure 5.4 in which

$$a_i = \frac{a}{3}, \ b_i = \frac{b}{3} \ (i = 1, 2, \cdots, 8),$$ (5.16)

and

$$\begin{aligned} &\alpha_1 = \alpha_5 = \alpha, \quad \alpha_2 = \alpha_4 = -\alpha_3 = \alpha, \quad \alpha_6 = \alpha_8 = -\alpha_7 = \alpha, \\ &\beta_1 = \beta_5 = \beta, \quad \beta_2 = \beta_4 = -\beta_3 = \beta, \quad \beta_6 = \beta_8 = -\beta_7 = \beta. \end{aligned}$$ (5.17)

The radius of the model, which inscribes the expanded assembly, can be obtained geometrically. To do so, let us consider an assembly consisting of nine large Bennett linkages, three in a row with three rows in total, Figure 5.5(a), where smaller intermediate Bennett linkages are not accounted for. The assembly satisfies Eqs (5.16) and (5.17).

Denote by l_0 the distance between two rows along the guideline, and by r_0 the radius of the circle that inscribes the assembly of Bennett linkages, both of which are indicated in Figures 5.5(a) and (b). To obtain l_0 and r_0 we need to consider the geometry of a typical Bennett linkage in three dimensional space first. Figure 5.6(a) shows a Bennett linkage ABCD being placed on a cylindrical surface in such a way that A, B, C and D are on the surface and line BD, colinear with a guideline, is along the longitudinal direction of the cylinder. The geometric parameters of Bennett linkage ABCD are

$$a_{AB} = a_{CD} = a, \ a_{AD} = a_{BC} = b, \ \alpha_{AB} = \alpha_{CD} = \alpha \text{ and } \alpha_{AD} = \alpha_{BC} = \beta.$$

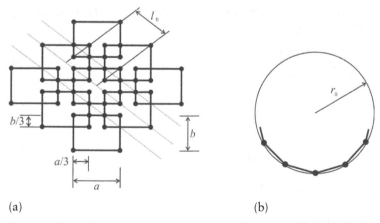

(a)   (b)

Figure 5.5 (a) Key geometrical parameters of an assembly and (b) the cross-section of the expanded profile.

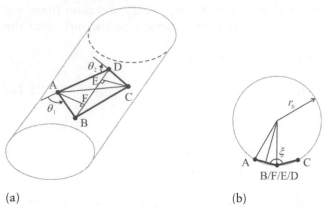

(a)   (b)

Figure 5.6 The geometry of a single Bennett linkage. (a) On the surface of a cylinder that inscribes the linkage and (b) the cross-sectional view.

Its revolute variables, θ_1 and θ_2 are marked in Figure 5.6(a), and are related by Eq. (2.27c). The cross-sectional view of the linkage on the cylinder is given in Figure 5.6(b), in which the angle between planes ABD and BCD is denoted by ξ. Take l_0 as the length BD. In $\triangle ABD$,

$$l_0 = BD = \sqrt{a^2 + b^2 + 2ab\cos\theta_1}. \tag{5.18}$$

Denote by r_0 the radius of this cylinder.

$$r_0 = \frac{AF}{2\cos\dfrac{\xi}{2}} = \frac{CE}{2\cos\dfrac{\xi}{2}}, \tag{5.19}$$

where both AF and CE are equal and perpendicular to BD. Considering $\triangle ABD$ or $\triangle CBD$, we have

$$AF = CE = \frac{ab\sin\theta_1}{\sqrt{a^2 + b^2 + 2ab\cos\theta_1}}. \tag{5.20}$$

Hence, from Figure 5.6(b),

$$\cos\xi = -\frac{(\cos\theta_1 + \cos\theta_2)(a^2 + b^2 + 2ab\cos\theta_1)}{ab\sin^2\theta_1} - 1. \tag{5.21}$$

Substituting Eqs (5.20) and (5.21) into Eq. (5.19) yields

$$r_0 = \frac{ab\sin^2\theta_1}{2(a^2 + b^2 + 2ab\cos\theta_1)}\sqrt{\frac{-2ab}{\cos\theta_1 + \cos\theta_2}}. \tag{5.22}$$

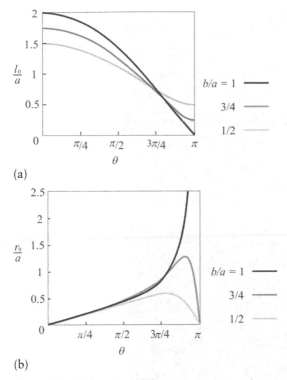

(a)

(b)

Figure 5.7 (a) l_0 vs θ_1 and (b) r_0/a vs θ_1 curves for a set of given b/a when $\alpha = \pi/3$.

For given α, l_0/a, r_0/a vs deployed angle θ_1 curves can be plotted using Eqs (5.18) and (5.22). Figure 5.7 shows such curves for $\alpha = \pi/3$.

A couple of observations can be made from the diagram. The first is when $b/a = 1$, i.e. the Bennett linkage is equilateral. The assembly folds in two configurations which correspond to $\theta_1 = \pi$, $l_0 = 0$ and, $\theta_1 = 0$, $r_0 = 0$, respectively. The deployment sequence, shown in Figure 5.8(a), starts from a compact bundle when $r_0 = 0$, expands *circumferentially* to an intermediate arch profile and then folds flat with $l_0 = 0$. This makes it ideal as a frame for expandable shelter. For assemblies made from non-equilateral Bennett linkages, i.e. $b/a \neq 1$, when θ changes from 0 to π, l_0 decreases whereas r_0 increases from 0 to a maximum value and then falls back to 0 again. This leads to the second notable case when the twists are set to be

$$\alpha_1 = \alpha_5 = \alpha, \alpha_2 = \alpha_4 = -\alpha_3 = -\alpha, \alpha_6 = \alpha_8 = -\alpha_7 = -\alpha,$$
$$\beta_1 = \beta_5 = \alpha, \beta_2 = \beta_4 = -\beta_3 = -\beta, \beta_6 = \beta_8 = -\beta_7 = -\beta.$$

(5.23)

The assembly resumes a completely *flat* profile throughout folding and expansion as shown in Figure 5.8(b).

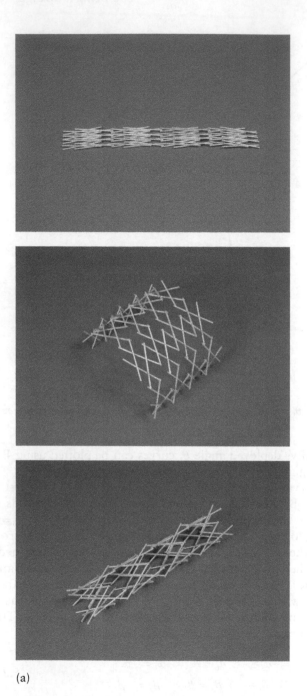

(a)

Figure 5.8 Motion sequence (a) an arch and (b) a grid with a flat profile.

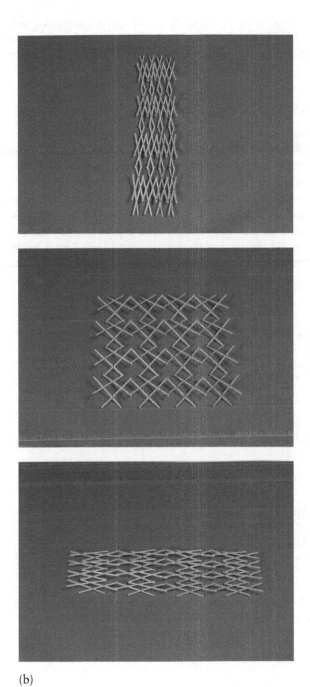

(b)

Figure 5.8 continued.

Note that some of lengths a_i, b_i $(i=1, 2,..., 8)$ can be negative, which also lead to mobile assemblies as long as Eqs (5.1) and (5.15) are satisfied. Some of these assemblies exhibit rather interesting profiles. Readers may refer to Chen and You (2008b) for details.

5.3 Multi-layer assemblies of Bennett linkages

The assembly of Bennett linkages presented so far does not have multiple layers. A typical portion of single-layer assembly of Bennett linkages surrounding a large Bennett linkage is redrawn as Figure 5.9(a). Again the twists are marked alongside the links. The guideline passes through A, E, G and C. In the expanded spiral profile, the top right Bennett linkage PVND and the bottom left Bennett linkage JBKS are roughly placed along the tangent of the cylindrical surface, with V and S pointing towards each other. The following proof shows that they can be connected by a bridging 4R loop RSTV to form the assembly shown in Figure 5.9(b) while retaining mobility.

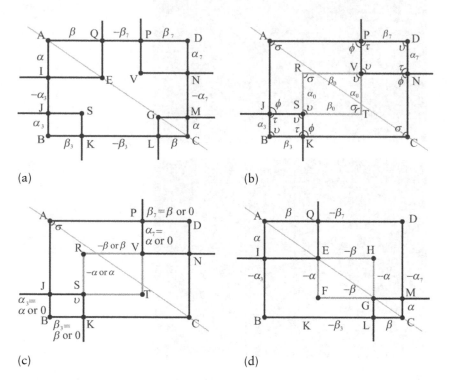

Figure 5.9 Formation of additional loops in a single-layer assembly. (a) A portion of a single-layer assembly; (b) formation of loop RSTV at a layer different from loop ABCD and (c) its corresponding twists; (d) formation of loop EFGH and its corresponding twists.

The revolute variables for each joint and twists for each segment of a link are given in Figure 5.9(b). Thus, the following relationships must hold.

In Bennett linkage ABCD,

$$\tan\frac{\pi-\upsilon}{2}\tan\frac{\pi-\sigma}{2}=\frac{\sin\frac{1}{2}(\beta+\alpha)}{\sin\frac{1}{2}(\beta-\alpha)};\tag{5.24}$$

JBKS,

$$\tan\frac{\pi-\upsilon}{2}\tan\frac{\pi-\tau}{2}=\frac{\sin\frac{1}{2}(\beta_3+\alpha_3)}{\sin\frac{1}{2}(\beta_3-\alpha_3)};\tag{5.25}$$

PVND,

$$\tan\frac{\pi-\upsilon}{2}\tan\frac{\pi-\tau}{2}=\frac{\sin\frac{1}{2}(\beta_7+\alpha_7)}{\sin\frac{1}{2}(\beta_7-\alpha_7)};\tag{5.26}$$

RSTV,

$$\tan\frac{\pi-\upsilon}{2}\tan\frac{\pi-\sigma}{2}=\frac{\sin\frac{1}{2}(\beta_0+\alpha_0)}{\sin\frac{1}{2}(\beta_0-\alpha_0)}.\tag{5.27}$$

In AJTP, twists become $\alpha_0+\alpha_7$, $\beta_0+\beta_3$, and variables are $\pi-\phi$, $\pi-\sigma$, thus,

$$\tan\frac{\pi-\phi}{2}\tan\frac{\pi-\sigma}{2}=\frac{\sin\frac{1}{2}((\beta_0+\beta_3)+(\alpha_0+\alpha_7))}{\sin\frac{1}{2}((\beta_0+\beta_3)-(\alpha_0+\alpha_7))}.\tag{5.28}$$

Similarly in RKCN with twists being $\alpha_0+\alpha_3$, $\beta_0+\beta_7$, and variables $\pi-\phi$, $\pi-\sigma$,

$$\tan\frac{\pi-\phi}{2}\tan\frac{\pi-\sigma}{2}=\frac{\sin\frac{1}{2}((\beta_0+\beta_7)+(\alpha_0+\alpha_3))}{\sin\frac{1}{2}((\beta_0+\beta_7)-(\alpha_0+\alpha_3))}.\tag{5.29}$$

From Eqs (5.24) to (5.29), the following relationships among the twists can be obtained:

$$\frac{\sin\frac{1}{2}(\beta+\alpha)}{\sin\frac{1}{2}(\beta-\alpha)}=\frac{\sin\frac{1}{2}(\beta_0+\alpha_0)}{\sin\frac{1}{2}(\beta_0-\alpha_0)}, \tag{5.30}$$

$$\frac{\sin\frac{1}{2}(\beta_3+\alpha_3)}{\sin\frac{1}{2}(\beta_3-\alpha_3)}=\frac{\sin\frac{1}{2}(\beta_7+\alpha_7)}{\sin\frac{1}{2}(\beta_7-\alpha_7)}, \tag{5.31}$$

$$\frac{\sin\frac{1}{2}((\beta_0+\beta_3)+(\alpha_0+\alpha_7))}{\sin\frac{1}{2}((\beta_0+\beta_3)-(\alpha_0+\alpha_7))}=\frac{\sin\frac{1}{2}((\beta_0+\beta_7)+(\alpha_0+\alpha_3))}{\sin\frac{1}{2}((\beta_0+\beta_7)-(\alpha_0+\alpha_3))} \tag{5.32}$$

and

$$\frac{\sin\frac{1}{2}(\beta_3+\alpha_3)}{\sin\frac{1}{2}(\beta_3-\alpha_3)}\cdot\frac{\sin\frac{1}{2}((\beta_0+\beta_7)+(\alpha_0+\alpha_3))}{\sin\frac{1}{2}((\beta_0+\beta_7)-(\alpha_0+\alpha_3))}=\frac{\sin\frac{1}{2}(\beta_0+\alpha_0)}{\sin\frac{1}{2}(\beta_0-\alpha_0)}. \tag{5.33}$$

Two sets of solutions emerge for Eqs (5.30) to (5.33) if α, $\beta\neq0$ or π, which are

$$\begin{aligned}\alpha_3=\alpha_7=\alpha, \quad &\alpha_0=-\alpha,\\ \beta_3=\beta_7=\beta, \quad &\beta_0=-\beta,\end{aligned} \tag{5.34}$$

and

$$\begin{aligned}\alpha_3=\alpha_7=0, \quad &\alpha_0=\alpha,\\ \beta_3=\beta_7=0, \quad &\beta_0=\beta.\end{aligned} \tag{5.35}$$

When α, $\beta\neq0$ or π, the assembly becomes a series of overlapped planar crossed isograms or planar parallelograms.

Denote the lengths of links for small Bennett linkages JBKS, PVND and RSTV as a_3, b_3, a_7, b_7 and a_0, b_0. Obviously, there must be

$$\frac{a_3}{b_3}=\frac{a_7}{b_7}=\frac{a_0}{b_0}=\frac{a}{b}, \tag{5.36}$$

and

$$a_3+a_0+a_7=a , \; b_3+b_0+b_7=b. \tag{5.37}$$

The solutions given in Eqs (5.34) and (5.35) indicate that, in order to connect the top right 4R loop PVND and the bottom left 4R loop JBKS, they must have either zero twists or twists identical to those of larger 4R loop. Both possibilities are shown in Figure 5.9(c) where twists are marked alongside corresponding links.

The connectivity of the top left 4R loop AIQE and the bottom right 4R loop GLCM along the other diagonal direction by an additional 4R loop EFGH, Figure 5.9(d), can also be examined using the similar analysis. It is found that connectivity is possible only in two cases. The first case is when loops AIQE and GLCM have the same twists as those of 4R loop ABCD, which are α and β. The twists of the connection 4R loop EFGH are $-\alpha$ and $-\beta$. The second case is when loops AIQE and GLCM have zero twists, i.e. both are planar 4R loops, and the twists of the connection 4R loop EFGH are α and β.

The above arrangements can be applied to the entire single-layer assembly shown in Figure 5.3(b), a portion of which is now redrawn in Figure 5.10(a), with the following conclusions:

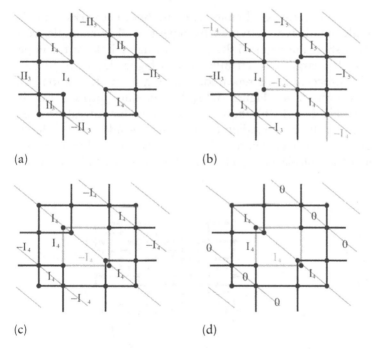

(a)

(b)

(c)

(d)

Figure 5.10 Construction of double-layer assembly of Bennett linkages. (a) A portion of single layer layout with guidelines. (b) Connections along the guidelines. Grey Bennett linkages are added which form the additional layer. Constructions in the other diagonal directions become possible provided that either (c) $II_3 = I_4 = II_5$ or (d) twists of II_3 and II_5 are zero. The newly added Bennett linkages are in grey and '0' indicates Bennett linkages with zero twists.

a   4R loops denoted by even number subscripts alongside the guidelines, e.g. I_2, I_4 and I_6, can always be connected by additional Bennett linkages of the same type but at a different layer forming a double-layer assembly, see Figure 5.10(b);

b   4R loops in the other diagonal direction, e.g. I_1, I_3, I_5 and I_7, can also be connected provided that either

- twists of two smaller loops must be identical to those of the larger loop that encircle them, e.g. to connect I_3 and I_5, both of them must be identical to I_4, see Figure 5.10(c), or

- twists of two smaller loops are zero, e.g. to connect I_3 and I_5, the twists of I_3 and I_5 are 0, see Figure 5.10(d).

Note that here we say that two Bennett linkages are *identical* if both linkages have the same twists whereas the link lengths, obtainable from Eq. (2.24), may be different.

The assemblies formed using the rules above will have double layers as the bridging 4R loops are at a level different from that of the single-layer assembly.

The solutions for forming double-layer mobile networks can be extended to build multi-layer networks by repetition. For example, consider a single unit shown in Figure 5.9(c), If the bridging links at the upper layer are extended, a larger Bennett linkage at a higher level can be connected to it. The process can be repeated, resulting in the formation of a mobile multi-layer mast, see Figure 5.11(a). The same process can be applied laterally. Figure 5.11(b) shows another possible lateral arrangement. Each of the 4R loops, including the ones whose sides have different shades, is a Bennett linkage. Figure 5.12(a) shows the deployment of a physical model of a multi-layer mobile assembly in which links are made of Al-alloy rods.

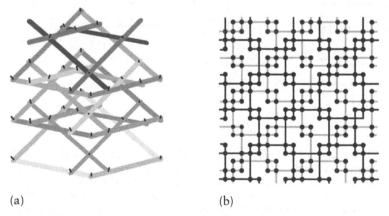

(a)   (b)

Figure 5.11 (a) A multi-layer mast and (b) another possible layout for assembly of Bennett linkages.

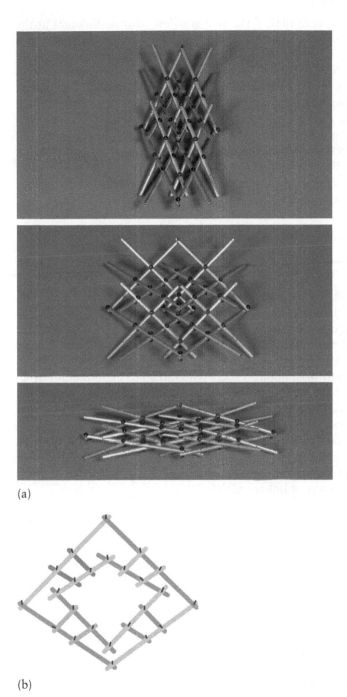

(a)

(b)

Figure 5.12 (a) Deployment sequence of a multi-layer network of Bennett linkages and (b) an assembly extracted from the multi-layer network which illustrates how two Bennett linkages can be connected while retaining mobility.

It is interesting to note that, if the two smallest Bennett linkages at the centre of the model were removed, the assembly would show the largest and second largest Bennett linkages connected by four smaller Bennett linkages on each side, Figure 5.12(b), forming a mobile assembly. This effectively shows that two Bennett linkages can also be connected together just like the connection of two planar four-bar linkages illustrated in Section 3.2.1. The difference is that smaller Bennett linkages have to be used to facilitate the connection instead of single pins in planar cases. The question of connecting of two Bennett linkages while retaining mobility was first raised by Baker and Hu (1986). A solution to this question can be extracted from the multi-layer assemblies just being created. Detailed proof can be found in Chen and You (2002) and Chen and Baker (2005).

5.4 Alternative form of Bennett linkage

5.4.1 The equilateral Bennett linkage and its alternative forms

The Bennett linkage discussed so far has its four links spanning the shortest distance between the axes of the neighbouring revolute joints. Although this enables us to uniquely describe the linkage mathematically, in a physical model constructed with these types of links, it is found that the linkage cannot be folded up completely in both directions linking two diagonal joints simultaneously. It duly disappoints readers who intend to build motion structures that fold to a compact bundle. However, in this section we demonstrate that modifications can be carried out to an equilateral Bennett linkage so that compact folding becomes possible (Chen and You, 2006).

To design for compact packaging, an equilateral Bennett linkage is drawn in three dimensions in Figure 5.13(a) with its joints marked with letters A, B, C and D, which correspond to 1, 2, 3 and 4 in Figure 5.1(a). So the lengths and twists of linkage satisfy

$$a_{AB} = a_{BC} = a_{CD} = a_{DA} = l, \tag{5.38}$$

$$\alpha_{AB} = \alpha_{CD} = \alpha,$$

$$\alpha_{BC} = \alpha_{DA} = \pi - \alpha, \tag{5.39}$$

due to Eq. (2.23). This linkage is symmetric both about the plane through AC and perpendicular to BD and about the plane through BD and perpendicular to AC even though lines AC and BD may not cross each other. The axes of revolute joints are marked at A, B, C and D by lines with arrows which give the positive directions of the axes.

Denote M and N as the respective middle points of BD and AC, see Figure 5.13(b). Obviously, $\triangle$ABD and $\triangle$CDB are isosceles and identical triangles due to Eq. (5.38). So are $\triangle$BCA and $\triangle$DAC. These lead to the

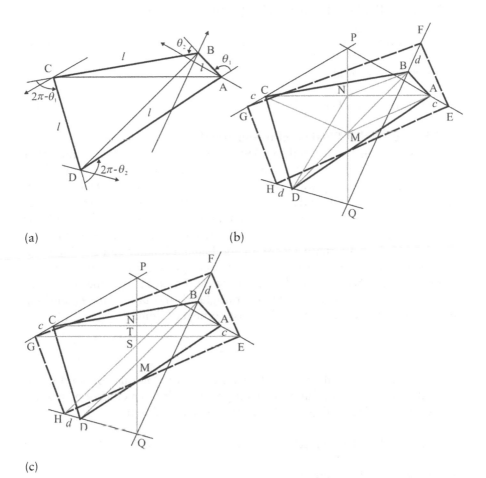

(a)

(b)

(c)

Figure 5.13 Equilateral Bennett linkage. Certain new lines are introduced in (a), (b) and (c) for derivation of compact folding and maximum expansion conditions.

conclusion that ΔAMC and ΔBND are both isosceles triangles. Hence MN is perpendicular to both AC and BD. Moreover, extensions of the axes of revolute joints must meet with the extension of MN at P and Q, respectively, due to symmetry.

Consider now four alternative connection points E, F, G and H along the extensions of the revolute axes AP, BQ, CP and DQ, respectively, Figure 5.13(b). To preserve symmetry, let $GC = AE = c$ and $BF = DH = d$. We have

$$EF^2 = l^2 + c^2 + d^2 - 2cd \cos(\pi - \alpha_{AB}), \tag{5.40a}$$

$$FG^2 = l^2 + c^2 + d^2 - 2cd \cos \alpha_{BC}. \tag{5.40b}$$

Substituting Eq. (5.39) into Eqs (5.40a) and (5.40b) gives

$$EF = FG = \sqrt{l^2 + c^2 + d^2 + 2cd\cos\alpha}. \tag{5.41}$$

Similarly, we can find

$$GH = HE = EF = FG, \tag{5.42}$$

which suggests that EFGH is also equilateral.

For any given Bennett linkage ABCD, Eqs (5.41) and (5.42) show that EF, FG, GH and HE have constant length provided that both c and d are given. They do not vary with the revolute variables θ_1 or θ_2. Thus, it is possible to replace EF, FG, GH and HE with bars connected by the revolute joints whose axes are along BF, CG, DH and AE, respectively. EFGH is therefore an *alternative form* of the Bennett linkage ABCD. For each given set of c and d, an alternative form for the Bennett linkage can be obtained.

When the linkage in the alternative form displaces, the distance between E and G varies. So does the distance between F and H. Assume that when the linkage is fully folded, deployment angles θ_1 and θ_2 become θ_{1f} and θ_{2f}, respectively. The condition for the most compact folding is

$$EG = FH = 0, \tag{5.43}$$

indicating that physically the mechanism becomes a bundle. Eq. (5.43) can be written in term of c, d and the deployment angles, which is done next.

Consider $\triangle ADC$ in Figure 5.13(b). It can be found that

$$AC^2 = AD^2 + CD^2 - 2AD \cdot CD\cos(\pi - \theta_2) = 2l^2(1 + \cos\theta_2). \tag{5.44}$$

Similarly, in $\triangle ABD$, there is

$$BD^2 = AB^2 + AD^2 - 2AB \cdot AD\cos(\pi - \theta_1) = 2l^2(1 + \cos\theta_1), \tag{5.45}$$

whereas in right-angled triangle $\triangle BCM$,

$$CM^2 = BC^2 - BM^2 = \frac{l^2}{2}(1 - \cos\theta_1). \tag{5.46}$$

Thus, from $\triangle AMC$,

$$\cos\angle AMC = 1 - 2\frac{1 + \cos\theta_2}{1 - \cos\theta_1}. \tag{5.47}$$

From quadrilateral PAMC where PA and PC are perpendicular to MA and MC, respectively, we have

$$\cos \angle APC = -\cos \angle AMC = 2\frac{1+\cos\theta_2}{1-\cos\theta_1} - 1, \tag{5.48}$$

because in $\triangle APC$,

$$AC^2 = 2PC^2(1-\cos\angle APC) = 4PC^2 \frac{-\cos\theta_1 - \cos\theta_2}{1-\cos\theta_1}. \tag{5.49}$$

Comparing Eqs (5.44) with (5.49) yields

$$PC^2 = -l^2 \frac{(1+\cos\theta_2)(1-\cos\theta_1)}{2(\cos\theta_2+\cos\theta_1)}. \tag{5.50}$$

Similarly it can be obtained that

$$QB^2 = -l^2 \frac{(1+\cos\theta_1)(1-\cos\theta_2)}{2(\cos\theta_2+\cos\theta_1)}. \tag{5.51}$$

In $\triangle EPG$, there is

$$EG^2 = 2(PC+c)^2(1-\cos\angle APC), \tag{5.52}$$

and similarly

$$FH^2 = 2(QB+d)^2(1-\cos\angle BQD). \tag{5.53}$$

In general, $\angle APC$ and $\angle BQD$ cannot reach zero simultaneously. Substituting Eqs (5.52) and (5.53) into (5.43), noting that Eq. (5.43) holds only when the linkage is fully folded, i.e. $\theta_1 = \theta_{1f}$ and $\theta_2 = \theta_{2f}$, we have

$$c = -PC = -l\sqrt{\frac{(1+\cos\theta_{2f})(1-\cos\theta_{1f})}{-2(\cos\theta_{2f}+\cos\theta_{1f})}}, \tag{5.54a}$$

$$d = -QB = -l\sqrt{\frac{(1-\cos\theta_{2f})(1+\cos\theta_{1f})}{-2(\cos\theta_{2f}+\cos\theta_{1f})}}. \tag{5.54b}$$

The above equations show how the values of c and d are related to the fully folded revolute angles θ_{1f} and θ_{2f}. Both values are negative, implying that E, F, G and H must locate within lines PA, QB, PC and QD, respectively, rather than being at their extensions. In fact, c and d can be determined graphically, as Eq. (5.54) simply indicates that E and G should move to a single point P, and F and H to Q, if the configuration shown Figure 5.13(b) represents the fully folded configuration of linkage EFGH.

Having obtained the linkage corresponding to the most efficient folding configuration, what is the form of Bennett linkage that covers the largest area? To answer this question, it is necessary to find out the geometrical condition relating to the maximum coverage.

Figure 5.13(c) shows the alternative form of the Bennett linkage EFGH. Due to symmetry, a line between E with G will intersect MN at T, and that between F and H will intersect MN at S. The projection of EFGH will cover a maximum area if

$$ST = 0 \tag{5.55}$$

when revolute angles reach θ_{1o} and θ_{2o}. This implies that EFGH is completely flattened to a rhombus.

Again, ST can be expressed in term of c, d and deployment angles. Based on Eqs (5.44) and (5.46),

$$MN^2 = CM^2 - CN^2 = CM^2 - \frac{AC^2}{4} = -\frac{l^2}{2}(\cos\theta_1 + \cos\theta_2). \tag{5.56}$$

Considering Eq. (5.48) gives

$$\sin\angle PGE = \sin(\frac{\pi}{2} - \frac{1}{2}\angle APC) = \frac{\cos\dfrac{\theta_2}{2}}{\sin\dfrac{\theta_1}{2}}.$$

So,

$$NT = c \cdot \sin\angle PGE = c \cdot \frac{\cos\dfrac{\theta_2}{2}}{\sin\dfrac{\theta_1}{2}}. \tag{5.57}$$

Similarly,

$$MS = d \cdot \sin\angle QFH = d \cdot \frac{\cos\dfrac{\theta_1}{2}}{\sin\dfrac{\theta_2}{2}}. \tag{5.58}$$

Considering Eqs (5.56), (5.57) and (5.58), ST can be written as

$$ST = MN - MS - NT = l\sqrt{-\frac{(\cos\theta_1 + \cos\theta_2)}{2}} - d\frac{\cos\dfrac{\theta_1}{2}}{\sin\dfrac{\theta_2}{2}} - c\frac{\cos\dfrac{\theta_2}{2}}{\sin\dfrac{\theta_1}{2}}. \tag{5.59}$$

When $\theta_1 = \theta_{1e}$ and $\theta_2 = \theta_{2e}$

$$ST = l\sqrt{-\frac{(\cos\theta_{1e} + \cos\theta_{2e})}{2} - d\frac{\cos\dfrac{\theta_{1e}}{2}}{\sin\dfrac{\theta_{2e}}{2}} - c\frac{\cos\dfrac{\theta_{2e}}{2}}{\sin\dfrac{\theta_{1e}}{2}}} = 0, \tag{5.60}$$

due to Eq. (5.55).

Parameters c and d that satisfy Eq. (5.60) corresponded to an alternative form that provides maximum coverage when fully expanded.

5.4.2 *Parametrical study*

Parameters c and d obtained from either Eqs (5.54) or (5.60) are functions of the dimensional parameters of the original equilateral Bennett linkage, α and l, and initial and final revolute variables θ_{1f}, θ_{2f}, θ_{1e} and θ_{2e}. Bear in mind that only two of the four revolute variables, one for fully folded and the other for the extended configurations, are independent because

$$\tan\frac{\theta_{1e}}{2}\tan\frac{\theta_{2e}}{2} = \tan\frac{\theta_{1f}}{2}\tan\frac{\theta_{2f}}{2} = \frac{1}{\cos\alpha}, \tag{5.61}$$

due to Eq. (2.28).

With a set of these parameters, we are able to obtain an alternative form that can have both compact folding and maximum coverage. In other words, c and d must satisfy both Eqs (5.54) and (5.60).

Substituting c and d obtained from Eqs (5.54a) and (5.54b) into Eq. (5.60) and then considering Eq. (5.61) give

$$\tan^4\alpha\tan^2\frac{\theta_{1e}}{2}\tan^2\frac{\theta_{1f}}{2} = \left(\sec^2\frac{\theta_{1e}}{2}\sec^2\frac{\theta_{1f}}{2} + \tan^2\alpha\right)^2. \tag{5.62}$$

If $0 \le \theta_{1f} \le \pi$, then $\pi \le \theta_{1e} \le 2\pi$. Eq. (5.62) becomes

$$-\tan^2\alpha\tan\frac{\theta_{1e}}{2}\tan\frac{\theta_{1f}}{2} = \sec^2\frac{\theta_{1e}}{2}\sec^2\frac{\theta_{1f}}{2} + \tan^2\alpha. \tag{5.63}$$

Should either θ_{1f} or θ_{1e} be predetermined, the other can be obtained from Eq. (5.63).

Solutions to Eq. (5.63) only exist when the value of α is in the range between arccos(1/3) and π−arccos(1/3), i.e. 70.53° to 109.47°. Within this range, the relationship between θ_{1f} and θ_{1e} for a set of given α is shown in Figure 5.14. Note that in most circumstances, each θ_{1f} corresponds to two values of θ_{1e}. This means that there are two possible expanded configurations in which the linkage in its alternative form can be flattened. For

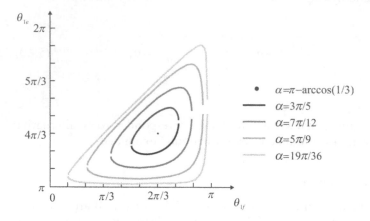

Figure 5.14 θ_{1f} vs θ_{1e} for a set of given α.

$\alpha < \arccos(1/3)$ or $\alpha > \pi - \arccos(1/3)$, there is no pair of θ_{1f} and θ_{1e} satisfying Eq. (5.63). Therefore the linkage is incapable of being flattened despite that it can be folded up compactly, or vice versa.

The actual side length of the alternative form of the Bennett linkage, L, can be obtained from Figure 5.13(b) as

$$L = \sqrt{l^2 + c^2 + d^2 + 2cd\cos\alpha}. \qquad (5.64)$$

Using c and d obtained from Eq. (5.54),

$$\frac{L}{l} = \sqrt{\frac{-2}{\cos\theta_{1f} + \cos\theta_{2f}}} = \frac{\sqrt{1 + \tan^2\dfrac{\theta_{1f}}{2}\cos^2\alpha}}{\sin\dfrac{\theta_{1f}}{2}\sin\alpha}. \qquad (5.65)$$

Denote by δ the angle between two adjacent sides of the alternative form of the Bennett linkage in its flattened configuration when $\theta_1 = \theta_{1e}$ and $\theta_2 = \theta_{2e}$. Thus, $\delta = \angle FGH$ when S and T in Figure 5.13(c) become one point. We have

$$\tan\frac{\delta}{2} = \frac{\dfrac{1}{2}FH}{\dfrac{1}{2}EG} = \frac{FH}{EG}.$$

Expressing FH and EG in terms of angles gives

$$\delta = 2\arctan\left(\frac{\sqrt{\dfrac{(1-\cos\theta_{2f})(1+\cos\theta_{1f})}{-2(\cos\theta_{2f}+\cos\theta_{1f})}}+\sqrt{\dfrac{(1-\cos\theta_{2e})(1+\cos\theta_{1e})}{-2(\cos\theta_{2e}+\cos\theta_{1e})}}}{\sqrt{\dfrac{(1+\cos\theta_{2f})(1-\cos\theta_{1f})}{-2(\cos\theta_{2f}+\cos\theta_{1f})}}+\sqrt{\dfrac{(1+\cos\theta_{2e})(1-\cos\theta_{1e})}{-2(\cos\theta_{2e}+\cos\theta_{1e})}}}\right.$$

$$\left.\sqrt{\frac{1-\cos\theta_{1e}}{1-\cos\theta_{2e}}}\right). \tag{5.66}$$

Similar to relationship between θ_{1e} and θ_{1f}, there are two values of δ for each θ_{1f}.

The flattened configuration typically has a rhombus shape. Among all the rhombuses with the same side length the square has the largest area, i.e.

$$\delta = \frac{\pi}{2}. \tag{5.67}$$

Substituting Eq. (5.66) into Eq. (5.67) gives

$$\frac{\sqrt{\dfrac{(1-\cos\theta_{2f})(1+\cos\theta_{1f})}{-2(\cos\theta_{2f}+\cos\theta_{1f})}}+\sqrt{\dfrac{(1-\cos\theta_{2e})(1+\cos\theta_{1e})}{-2(\cos\theta_{2e}+\cos\theta_{1e})}}}{\sqrt{\dfrac{(1+\cos\theta_{2f})(1-\cos\theta_{1f})}{-2(\cos\theta_{2f}+\cos\theta_{1f})}}+\sqrt{\dfrac{(1+\cos\theta_{2e})(1-\cos\theta_{1e})}{-2(\cos\theta_{2e}+\cos\theta_{1e})}}}\cdot\sqrt{\frac{1-\cos\theta_{1e}}{1-\cos\theta_{2e}}}=1.$$

$$\tag{5.68}$$

Considering Eq. (5.61), Eq. (5.68) can be simplified as

$$\tan^2\frac{\theta_{1f}}{2}\sec^2\frac{\theta_{1e}}{2}=\tan^2\frac{\theta_{1e}}{2}+\sec^2\alpha. \tag{5.69}$$

So when θ_{1e} and θ_{1f} satisfy both of Eqs (5.63) and (5.69), the linkage based on the alternative form expands to a square. Solving both equations simultaneously, we obtain,

$$\sec^2\theta_{1f}\left(\tan^2\frac{\theta_{1f}}{2}-1\right)=\tan^2\alpha, \tag{5.70}$$

$$\theta_{1e}=2\theta_{1f}. \tag{5.71}$$

Moreover, for any square fully deployed configuration, we always have

$$L=\sqrt{2}l, \tag{5.72}$$

due to Eqs (5.61), (5.65) and (5.70).

Finally, it should be pointed out that normally, for a given α, there are two sets of θ_{1e} and θ_{1f}, in which configuration of the corresponding alternative form of Bennett linkage is square. However, when

$$\alpha = \arccos\frac{1}{3} \text{ or } \pi - \arccos\frac{1}{3},$$

there is only one solution,

$$\theta_f = \frac{2}{3}\pi \text{ and } \theta_e = \frac{4}{3}\pi.$$

In this case,

$$c = d = -\frac{\sqrt{6}}{4}l, \ L = \sqrt{2}l.$$

5.5 Physical model of alternative form of Bennett linkage

According to Crawford *et al.* (1973), for a close loop consisting of n links with identical cross-section, the cross-section of the whole assembly in the folded configuration should be n regular polygon if it can be packaged most compactly. So for the alternative form of Bennett linkage, each link could be made of four rods with square cross-section. The cross-section of the folded linkage will then be a square. A model is shown in Figure 5.15, which is defined by three design parameters: corner angle ω, tilting angle λ and rod length L. The relationship linking the design parameters to the geometric parameters of the alternative form of the Bennett linkage is derived next.

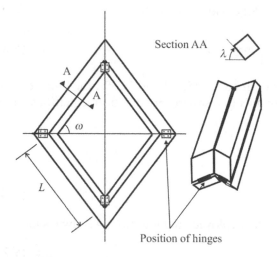

Section AA

Position of hinges

Figure 5.15 The alternative form of Bennett linkage made from square cross-section bars in the deployed and folded configurations.

Figure 5.16(a) shows an expanded linkage EFGH which is the alternative form of the Bennett linkage. The linkage has reached its maximum expansion and thus, EFGH becomes a plane rhombus. For simplification in description, let us define a coordinate system where axis x passes

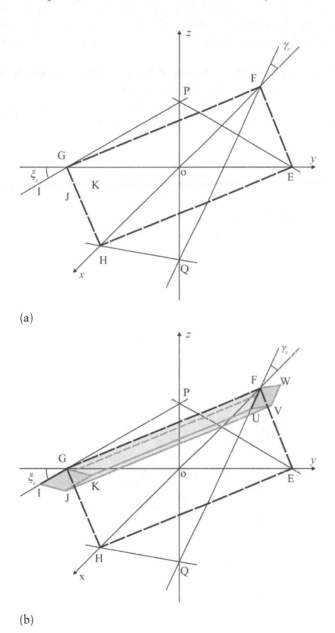

(a)

(b)

Figure 5.16 (a) The alternative form of Bennett linkage with (b) a rod of square cross-section used as link.

through FH and axis y through EG. The linkage EFGH is therefore symmetric about both xoz and yoz planes where o is the centre of EFGH and axis z is perpendicular to plane xoy. Now introduce a square bar, shown in light grey colour in Figure 5.16(b), to replace link FG in such a way that one of the edges of the square bar lies along FG. The bar is terminated by planes GIJK and FUVW, created by slicing the bar by the plane yoz and xoz, respectively. This bar is one of the rods in the model shown in Figure 5.15.

An enlarged diagram of bar FG is shown in Figure 5.17(a), in which GI and FU are the axes of the revolute joints. Plane $x'o'y'$ is through JV and parallel to plane xoy. A typical square cross-section is marked as RXYZ. The projection of the bar is given in Figure 5.17(b), in which X′, R′ and Z′,

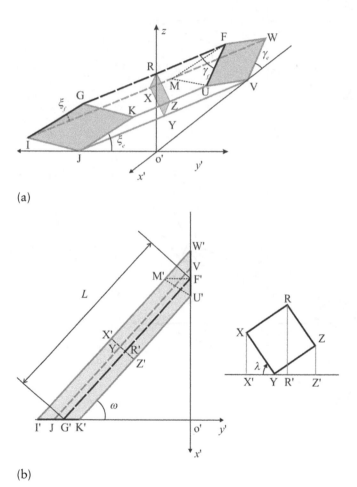

(a)

(b)

Figure 5.17 The geometry of the square cross-section bar. (a) In 3D and (b) projection on the plane $x'o'y'$ and the cross-section RXYZ.

etc. represent the projection of X, R and Z on the plane x'o'y', etc. The design parameters ω, λ and L are clearly shown in Figure 5.17(b).

Assume

$$\overline{RX} = \overline{XY} = \overline{YZ} = \overline{ZR} = 1.$$

Geometrically the following relationships can be obtained from Figure 5.17(b).

$$\overline{X'Y} = \overline{R'Z'} = 1 \cdot \cos \lambda,$$

$$\overline{X'R'} = \overline{YZ'} = 1 \cdot \sin \lambda,$$

$$\overline{X'Z'} = 1 \cdot (\sin \lambda + \cos \lambda),$$

$$\overline{II'} = \overline{XX'} = \overline{WW'} = 1 \cdot \sin \lambda,$$

$$\overline{KK'} = \overline{ZZ'} = \overline{UU'} = 1 \cdot \cos \lambda,$$

$$\overline{F'U'} = \overline{VW'} = \frac{\overline{R'Z'}}{\cos \omega} = \frac{\cos \lambda}{\cos \omega},$$

$$\overline{F'W'} = \overline{U'V} = \frac{\overline{X'R'}}{\cos \omega} = \frac{\sin \lambda}{\cos \omega},$$

$$\overline{G'K'} = \overline{I'J} = \frac{\overline{R'Z'}}{\sin \omega} = \frac{\cos \lambda}{\sin \omega},$$

$$\overline{G'I'} = \overline{JK'} = \frac{\overline{X'R'}}{\sin \omega} = \frac{\sin \lambda}{\sin \omega}.$$

So we have

$$\overline{GI} = \overline{JK} = \sqrt{\overline{JK'}^2 + \overline{KK'}^2} = \sqrt{\frac{\sin^2 \lambda}{\sin^2 \omega} + \cos^2 \lambda}, \qquad (5.73)$$

$$\overline{FU} = \overline{VW} = \sqrt{\overline{VW'}^2 + \overline{WW'}^2} = \sqrt{\frac{\cos^2 \lambda}{\cos^2 \omega} + \sin^2 \lambda}. \qquad (5.74)$$

Now draw line FM that is parallel to GI and crosses IW at M. Thus,

$$\angle MFU = \alpha,$$

and

$$\cos \alpha = \frac{\overline{FM}^2 + \overline{FU}^2 - \overline{UM}^2}{2 \cdot \overline{FM} \cdot \overline{FU}}. \qquad (5.75)$$

Because FM // GI and FG // MI, we have

$$\overline{FM} = \overline{GI} = \sqrt{\frac{\sin^2 \lambda}{\sin^2 \omega} + \cos^2 \lambda},$$ (5.76)

$$\overline{F'M'} = \overline{G'I'} = \frac{\sin \lambda}{\sin \omega},$$ (5.77)

and F'M' ⊥ W'U'. Then,

$$\overline{U'M'}^2 = \overline{F'M'}^2 + \overline{F'U'}^2,$$ (5.78)

and

$$\overline{UM}^2 = \overline{U'M'}^2 + \left(\overline{MM'} - \overline{UU'}\right)^2.$$ (5.79)

Substituting Eqs (5.74), (5.76) and (5.79) into (5.75), there is

$$\cos \alpha = \pm \frac{\sin 2\lambda \sin 2\omega}{\sqrt{\sin^2 2\lambda \sin^2 2\omega + 8\left(1 - \cos 2\lambda \cos 2\omega\right)}}.$$ (5.80)

Eq. (5.80) is the relationship between design parameters ω, λ and twist α of the original Bennett linkage, which is plotted in Figure 5.18 for a set of given α. It is interesting to note that, for $0 \le \lambda \le \pi/2$ and $0 \le \omega \le \pi/2$, the range of α is between arccos(1/3) and π−arccos(1/3), which is the same as that obtained from Eq. (5.63).

Our next step is to obtain the relationship among λ, ω and revolute variables θ_{1f}, θ_{2f}, θ_{1e} and θ_{2e}. Apply Eq. (5.48) to the expanded configuration,

$$\cos \angle EPG = \cos \angle APC = 2\frac{1 + \cos \theta_{2e}}{1 - \cos \theta_{1e}} - 1.$$ (5.81a)

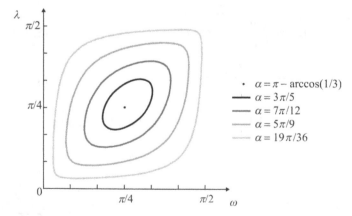

Figure 5.18 λ vs ω for a set of given α.

Similarly,

$$\cos \angle FQH = \cos \angle BQD = 2\frac{1+\cos\theta_{1e}}{1-\cos\theta_{2e}} - 1. \tag{5.81b}$$

From Figure 5.16(a),

$$\angle EPG = \pi - 2\xi_e \quad \text{and} \quad \angle FQH = \pi - 2\gamma_e. \tag{5.82a, b}$$

From Figure 5.17(a),

$$\tan\xi_e = \frac{\overline{KK'}}{\overline{JK'}} = \frac{\cos\lambda}{\sin\lambda/\sin\omega},$$

$$\tan\gamma_e = \frac{\overline{WW'}}{\overline{VW'}} = \frac{\sin\lambda}{\cos\lambda/\cos\omega}. \tag{5.83a, b}$$

From Eqs (5.81) to (5.83), it can be obtained that

$$\cos\theta_{1e} = \frac{(\cos 4\lambda - 1) + (\cos 4\lambda \cos 2\omega - 4\cos 2\lambda + 3\cos 2\omega)}{4(1-\cos 2\omega \cos 2\lambda)}, \tag{5.84a}$$

and

$$\cos\theta_{2e} = \frac{(\cos 4\lambda - 1) - (\cos 4\lambda \cos 2\omega - 4\cos 2\lambda + 3\cos 2\omega)}{4(1-\cos 2\omega \cos 2\lambda)}. \tag{5.84b}$$

When the linkage is folded up, EFGH becomes a bundle along ST, so do the rods with the square cross. Bar EF is shown in Figure 5.19 in with light grey colour.

Similarly to Eq. (5.81), in folded configuration, we have

$$\cos \angle ATC = 2\frac{1+\cos\theta_{2f}}{1-\cos\theta_{1f}} - 1, \tag{5.85a}$$

$$\cos \angle BSD = 2\frac{1+\cos\theta_{1f}}{1-\cos\theta_{2f}} - 1. \tag{5.85b}$$

From Figure 5.19,

$$\angle ATC = 2\xi_f \quad \text{and} \quad \angle BSD = 2\gamma_f. \tag{5.86a, b}$$

From Figure 5.17(a),

$$\tan\xi_f = \frac{\overline{RX}}{\overline{XI}-\overline{RG}} = \frac{\overline{RX}}{\overline{R'X'}/\tan\omega} = \frac{\tan\omega}{\sin\lambda}, \tag{5.87a}$$

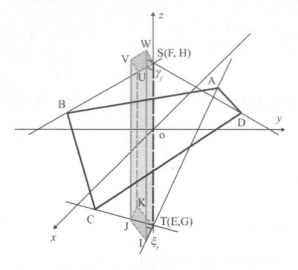

Figure 5.19 The alternative form of Bennett linkage with square cross-section bars in folded configuration.

$$\tan \gamma_f = \frac{\overline{RZ}}{\overline{RF} - \overline{UZ}} = \frac{\overline{RZ}}{\overline{R'Z'} \tan \omega} = \frac{1}{\cos \lambda \tan \omega}. \tag{5.87b}$$

Consider Eqs (5.85), (5.86) and (5.87) together, it can be found that

$$\cos \theta_{1f} = \frac{(\cos 4\omega - 1) + (\cos 4\omega \cos 2\lambda - 4\cos 2\omega + 3\cos 2\lambda)}{4(1 - \cos 2\omega \cos 2\lambda)}, \tag{5.88a}$$

$$\cos \theta_{2f} = \frac{(\cos 4\omega - 1) - (\cos 4\omega \cos 2\lambda - 4\cos 2\omega + 3\cos 2\lambda)}{4(1 - \cos 2\omega \cos 2\lambda)}. \tag{5.88b}$$

The relationship among λ, ω and L/l, c/l, d/l can be obtained as follows. First, substitute Eq. (5.88) into Eq. (5.65). There is

$$\frac{L}{l} = 2\sqrt{\frac{1 - \cos 2\omega \cos 2\lambda}{1 - \cos 4\omega}}. \tag{5.89}$$

When the expended configuration of the alternative form is square, i.e.

$$\omega = \frac{\pi}{4},$$

Eq. (5.89) gives

$$\frac{L}{l} = \sqrt{2},$$

which is the same as Eq. (5.72).

Then substituting Eq. (5.88) into Eq. (5.54) yields

$$\frac{c}{l} = -\frac{\cos\lambda\sin^2\omega}{\cos\omega}\sqrt{\frac{1-\sin^2\omega\sin^2\lambda}{\sin^2\lambda\cos^2\omega+\sin^2\omega\cos^2\lambda}}, \tag{5.90a}$$

$$\frac{d}{l} = -\frac{\sin\lambda\cos^2\omega}{\sin\omega}\sqrt{\frac{1-\cos^2\omega\cos^2\lambda}{\sin^2\lambda\cos^2\omega+\sin^2\omega\cos^2\lambda}}. \tag{5.90b}$$

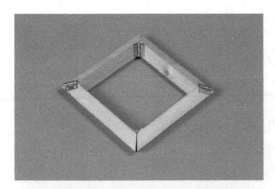

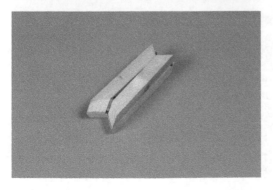

Figure 5.20 Model that $\lambda = \pi/6$ and $\omega = \pi/4$.

Obviously, α, θ_{1e}, θ_{2e}, θ_{1f} and θ_{2f} from Eqs (5.80), (5.84) and (5.88) satisfy Eq. (5.62).

Eqs (5.80), (5.89) and (5.90) relate the geometric parameters of an alternative form of Bennett linkage to the design parameters, based on which a model consisting of four square cross-section rods can be installed together as shown in Figure 5.15. The linkage has mobility one and can be folded into a compact bundle and deployed into a rhombus. In practice, it is not necessary to know the characteristics of the original Bennett linkage l, α, and extended distance on the axes of joints c (or d).

Figure 5.21 Model that $\lambda = \pi/4$ and $\omega = \pi/3$.

The design work can be done based on the values of L, λ and ω, though it is much more convenient to use l, α and c (or d) when analysing the kinematic behaviour of the linkage such as θ_{1f}, θ_{2f}, θ_{1e} and θ_{2e}. Several models have been made which are shown in Figures 5.20 and 5.21 (Chen and You, 2007a).

5.6 Assemblies of alternative form linkages

The Bennett linkage in its alternative forms can also be used to construct large mobile assemblies with the same layout as that for the linkage in its original form.

The layout for a single layer assembly of Bennett linkages is shown in Figure 5.22, which is similar to that shown in Figure 5.3(b) but with two crucial modifications. First, all the large rectangles are replaced by squares, since the alternative forms presented in the previous section use only equilateral Bennett linkages with twists being α and $\pi-\alpha$. Second, all of the squares are the same size because it is likely to be preferred that the assembly folds to a compact bundle with all the rods neatly packed together with the same height. Note that now each of the squares in the layout still represents a Bennett linkage but in its alternative form.

The same linkages as those shown in Figures 5.20 or 5.21 are used for the 4R loops represented by large squares in the layout. The models in Figures 5.20 or 5.21, have their revolute joints made from door hinges, which are placed at the top face of the models. The same is done to linkages in the layout. These linkages are to be connected to a neighbouring one at each of its corners according to the layout. To enable such connections, the sharp

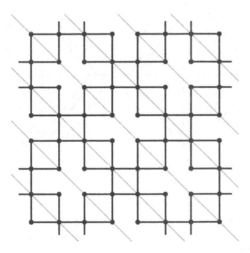

Figure 5.22 Layout of a single-layer assembly consisting of equilateral Bennett linkages.

corner tips of each linkage have to be removed, Figures 5.23(a) and (b). Two more revolute joints, placed at the rear face of the linkage, are added at each corner for the connection with neighbours, Figure 5.23(c). The additional hinges at each corner, together with original corner joints, form four revolute joints of each small 4R loop in the layout, represented by small squares. The folding sequence of a model is given in Figure 5.24(a), accompanied by a close-up of the corner joints in Figure 5.24(b). This particular model expands to a flat profile and can be packaged to a compact bundle.

Assemblies that expand to a cylindrical profile can also be produced. Similar to the single layer assembly of the original Bennett linkages, the Bennett linkages along the guideline expands to a flat profile. Non-zero curvature is allowed only in the direction perpendicular to the guidelines. Figures 5.25 and 5.26 show computer simulations of the expansion of two models. The first model forms the top part of a hexagonal prism and the second forms a square prism. Details of how to design an assembly with cylindrical profile can be found in Chen and You (2007c) as well as Tian and Chen (2010).

5.7 Applications

One obvious application of the spatial assembly of Bennett linkages is as the frame structure of rapidly erectable shelters because of its ability to form a curved profile. One can take a strip out of the single-layer layout of Figure 5.3(b) in the direction perpendicular to the guidelines. The resulting

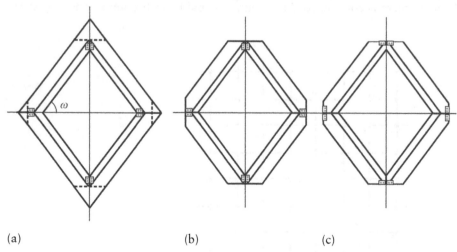

(a)   (b)   (c)

Figure 5.23 Modifying the Bennett linkage in its alternative form for construction of a flat-profile assembly. (a) A single Bennett linkage in its alternative form with sharp corners; (b) sharp corners removed and (c) rear view of the same linkage showing two additional joints for connection with neighbouring linkages.

(a)

(b)

Figure 5.24 (a) Folding sequence of a single layer assembly of Bennett linkages in their alternative forms and (b) the front and rear of a corner joint.

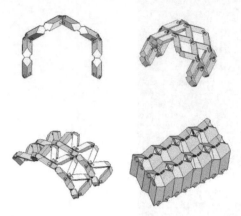

Figure 5.25 Folding simulation of a model forming the top part of a hexagonal prism.

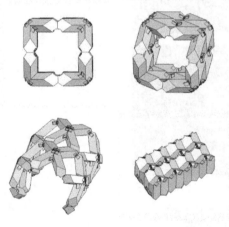

Figure 5.26 Folding simulation of a model forming a square prism.

assembly could have a layout of Figure 5.27(a) or Figure 5.27(b). The selection is primarily governed by the package length if original Bennett linkages are used. Figure 5.27(c) is the model constructed using the layout of Figure 5.27(b).

Another issue associated with the assembly formed from the original Bennett linkage is the joint size which we have not considered until now. In the models displayed in Figures 5.6 and 5.8, all of the rods are placed along the links given in the layout and the revolute joint consists of a pin

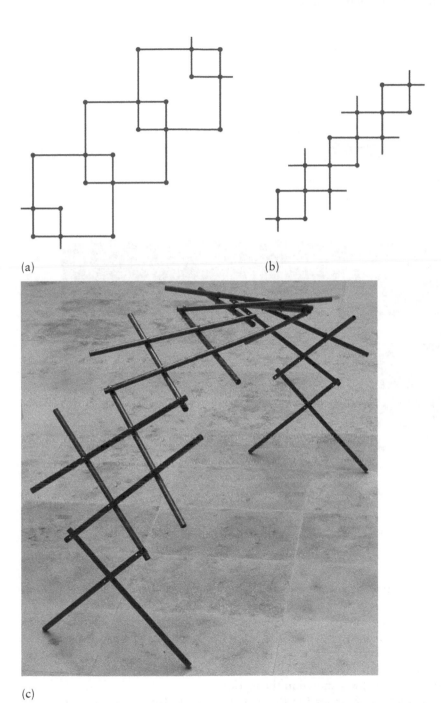

(a)

(b)

(c)

Figure 5.27 (a) and (b) Arch frames obtained by taking a strip out of a single layer assembly of Bennett linkages and (c) a model based on (b).

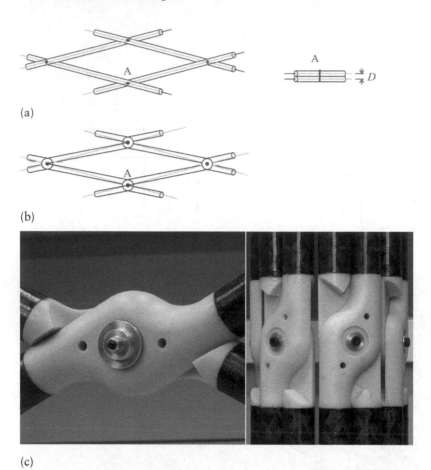

(a)

(b)

(c)

Figure 5.28 (a) Error caused assembling straight rods with a single pin connection
as revolute joints, (b) a new joint design that removes the error caused
by a single pin connection and (c) a joint model using the new joint
design.

through a pair of circular rods. Though it is effective in demonstration of
the concept, it creates an error at the joint because the axes of the rods are
at a distance D apart where D is the diameter of the rods, Figure 5.28(a),
which would have rendered the model unworkable if both the rods and
pins were sufficiently stiff. This error can be avoided by using a new type
of joint design, Figure 5.28(b) and (c). A drawback is that the rods are
slightly off the lines shown in the layout.

No such issue arises for the assembly made from Bennett linkages in
their alternative forms as the assemblies are designed with revolute joints
being precisely placed.

6 Spatial motion structures based on Bricard linkages

DOI: 10.1201/9781482266610-6

6.1 Threefold-symmetric Bricard linkages and its assemblies

The Bricard linkages reviewed in Section 2.3.5 are the only $6R$ overconstrained linkages that are not derived from $4R$, $5R$ or other $6R$ linkages. Of a total of six types of Bricard linkages, the most suitable ones for the purpose of constructing motion structures is the threefold-symmetric Bricard linkage, obtained by combining the general plane-symmetric and trihedral Bricard linkages. The geometric parameters of the linkage satisfy the following conditions.

$$a_{12} = a_{23} = a_{34} = a_{45} = a_{56} = a_{61} = l,$$

$$\alpha_{12} = \alpha_{34} = \alpha_{56} = \alpha \,, \ \alpha_{23} = \alpha_{45} = \alpha_{61} = 2\pi - \alpha, \tag{6.1}$$

$$R_i = 0 \ (i = 1, 2, \cdots, 6).$$

The linkage has threefold rotational symmetry and also three planes of symmetry, hence its name. The configuration of this linkage is shown in Figure 6.1. It is easy to see that threefold-symmetric Bricard linkages form a subset of the set of plane-symmetric Bricard linkages. If, as a further specialisation, $\pi/2$ is selected as the angle α, then relationships given by Eq. (6.1) satisfy the geometrical conditions of trihedral Bricard linkages (Baker, 1980). Therefore, in this case, a threefold-symmetric Bricard linkage is also a trihedral Bricard linkage.

The linkage is mobile because of the plane-symmetric property guarantees it. It is however yet to know whether the number of degrees of freedom increases with an increase in the degree of symmetry. Because of threefold symmetry, the six revolute variables must satisfy the following conditions.

$$\theta_1 = \theta_3 = \theta_5,$$

$$\theta_2 = \theta_4 = \theta_6. \tag{6.2}$$

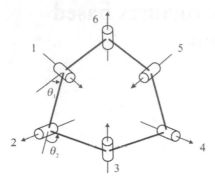

Figure 6.1 A threefold-symmetric Bricard linkage.

Since we have a six-link, single-loop linkage ($n = 6$), the closure condition (2.17) takes the form

$$[T_{61}][T_{56}][T_{45}][T_{34}][T_{23}][T_{12}] = [I],$$ (6.3)

or

$$[T_{34}][T_{23}][T_{12}] = [T_{45}]^{-1}[T_{56}]^{-1}[T_{61}]^{-1} = = [T_{54}][T_{65}][T_{16}].$$ (6.4)

Considering Eqs (6.1) and (6.2), the closure equation of the threefold-symmetric linkage can be extrapolated from Eq. (6.4), which is

$$\cos^2 \alpha \cos \theta_1 - \cos^2 \alpha \cos \theta_1 \cos \theta_2 - \cos^2 \alpha + \cos^2 \alpha \cos \theta_2$$
$$+ 2 \cos \alpha \sin \theta_1 \sin \theta_2 - \cos \theta_1 - \cos \theta_1 \cos \theta_2 - \cos \theta_2 = 0.$$ (6.5)

Eqs (6.2) and (6.5) form a set of independent closure equations for this 6R linkage. For any given α ($0 \leq \alpha \leq \pi$), Eq. (6.5) represents the input–output equation of the linkage. It is apparent that Eq. (6.5) is symmetric in θ_1 and θ_2, for the equation remains the same if these two variables are swapped. Therefore, one of the variables, either θ_1 or θ_2, can be chosen to be the input and the other can be obtained as the output. Figure 6.2 shows the input–output curve determined by Eq. (6.5). It is periodic and the periods for both θ_1 and θ_2 are 2π.

A number of distinctive features of the threefold-symmetric Bricard linkage with any twist α can be summarised from Figure 6.2. First of all, it shows that only one of six revolute variables can be free. Thus, in general, this threefold-symmetric Bricard linkage has mobility one. Second, the linkage with twist α behaves the same as that whose twist is $\pi - \alpha$. Third, all the input–output curves pass through the points (0, $-2\pi/3$), (0, $2\pi/3$), ($-2\pi/3$, 0) and ($2\pi/3$, 0), regardless of the value of α.

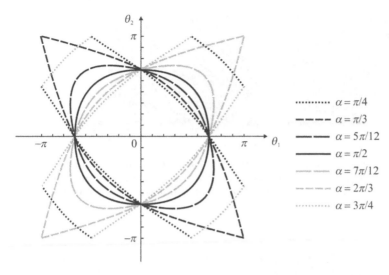

Figure 6.2 θ_2 versus θ_1 curves for the threefold-symmetric linkage.

This means that all of the threefold-symmetric Bricard linkages can be flattened to form a planar equilateral triangle whose side length is $2l$. Additionally, when $0 \leq \alpha \leq \pi/3$ or $2\pi/3 < \alpha \leq \pi$, the movement of the linkages is not continuous. It has been found by experiment that the linkage is physically blocked in the positions when all the links are crossed in the centre when either θ_1 or θ_2 reaches π or $-\pi$. Figure 6.3 shows two configurations of a model with $\alpha = \pi/4$. The input–output curve forms a closed loop when $\pi/3 \leq \alpha \leq 2\pi/3$ and thus the linkage keeps moving continuously. A model with $\alpha = 5\pi/12$ is shown in Figure 6.4, with a continuous range of movement.

Now focus on a particular set of input–output curves in Figure 6.2 when $\alpha = \pi/3$ or $2\pi/3$. Both θ_1 and θ_2 reach π or $-\pi$ simultaneously, which correspond to the configurations of the most compact folding where all of the links fold to a bundle. On the other hand, when $\theta_1 = 0$, $\theta_2 = 2\pi/3$ (or $-2\pi/3$), or vice versa, the linkage forms a plane equilateral triangle, in accordance to the configuration of maximum expansion. The motion sequence of a model with $\alpha = \pi/3$ is shown in Figure 6.5, which confirms the above findings. Because this particular threefold-symmetric Bricard linkage with twist of $\pi/3$ or $2\pi/3$ can achieve both compact folding and maximum expansion, it is an ideal building block for construction of large mobile assemblies, which will be discussed next.

The threefold-symmetric Bricard linkages with twist $\pi/3$ or $2\pi/3$ have the same behaviour and therefore only $\alpha = \pi/3$ is considered hereafter. This particular linkage can be represented by the schematic diagram shown in Figure 6.6(a), in which the hinge connecting the ends of the links is

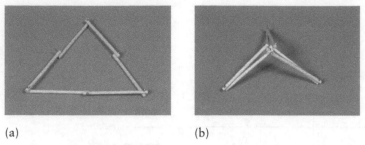

(a)   (b)

Figure 6.3 Motion sequence of a threefold-symmetric Bricard linkage with $\alpha = \pi/4$. (a) The configuration of planar equilateral triangle and (b) the configuration in which the movement of linkage is physically blocked.

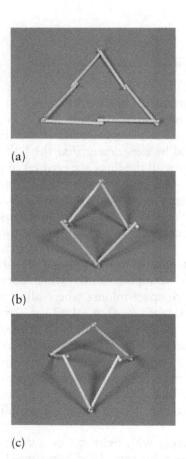

(a)

(b)

(c)

Figure 6.4 Motion sequence of a threefold-symmetric Bricard linkage with $\alpha = 5\pi/12$. (a) The configuration of planar equilateral triangle, (b) and (c) the configurations during the process of movement.

represented by a hollow circle. When the links are extended, which will be the case when the linkage is connected with its neighbours, the hinges may appear in the middle of a link. In this circumstance the hinge is represented by a dot. A pair of links connected by a hinge in the middle, Figure 6.6(b),

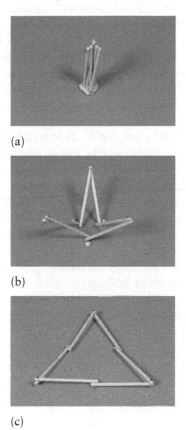

(a)

(b)

(c)

Figure 6.5 Motion sequence of a threefold-symmetric Bricard linkage with $\alpha = \pi/3$. The first image corresponds to the compact folding whereas the last is the maximum expansion.

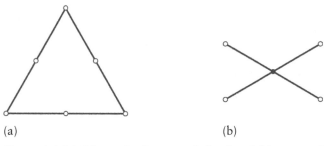

(a)   (b)

Figure 6.6 (a) Schematic diagram of the threefold-symmetric Bricard linkage and (b) a pair of links connected by a hinge in the middle.

is referred to as a pair of crossbars. It behaves like a pair of scissors. But it is different from the scissor-like elements introduced in Chapters 3 and 4 for the axes of the end and mid hinges of a crossbar may not be parallel.

The basic element for building a large mobile assembly is formed by having three pairs of crossbars with identical length in a closed chain with a threefold-symmetric Bricard linkage with $\alpha=\pi/3$ at the centre. There exist two possible arrangements of twists for a pair of crossbars, namely types I and II as shown in Figures 6.7(a) and (b), in which $\alpha=\pi/3$ and $\beta=2\pi-\pi/3$. There are in total four possible combinations of three pairs of crossbars, I-I-I, I-I-II, I-II-II and II-II-II, one of which is shown in Figure 6.7(c). It is possible to determine by experiments whether an assembly composed of these basic elements will retain the motion character of the threefold-symmetric Bricard linkage.

Consider a simple connection of two basic elements as shown in Figure 6.8. The marked pair of crossbars at the centre could be either types I or II. Figure 6.9 is a model where the intermediate pair of crossbars is a type I pair. It is obvious that two elements work well together. So the arrangement of type I can form a mobile assembly. In Figure 6.10, two elements are connected with a type II pair, but this assembly cannot be fully folded. The links of each side physically block each other in the folding process, Figure 6.10(b). As the result, we conclude that a motion structure cannot be connected with a type II pair. Thus, for the basic elements to be connected as shown in Figure 6.8, only arrangement I-I-I can be used in order to retain mobility. A typical portion of such an assembly is shown in Figure 6.11 and a physical model in Figure 6.12.

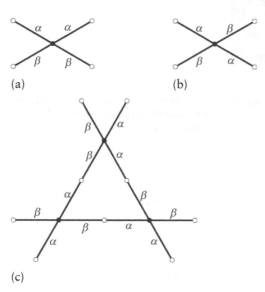

(a)

(b)

(c)

Figure 6.7 Construction of a basic element. (a) Type I and (b) Type II of the crossbars; (c) one of the arrangements: I-I-II.

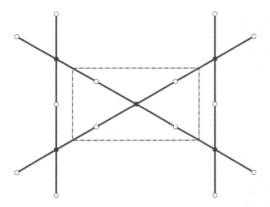

Figure 6.8 Connectivity of two basic elements.

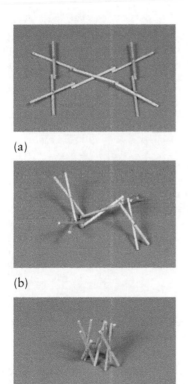

(a)

(b)

(c)

Figure 6.9 Model showing connectivity of two basic elements with a type I pair. (a) Fully expanded, (b) during deployment and (c) close to being fully folded.

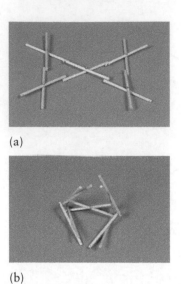

(a)

(b)

Figure 6.10 Model showing connectivity of two basic elements with a type II pair. (a) Fully expanded and (b) during deployment.

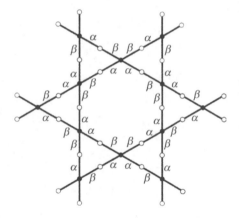

Figure 6.11 Portion of an assembly of threefold-symmetric Bricard linkages.

One of the issues associated with the threefold-symmetric Bricard linkage with twists of $\pi/3$ or $2\pi/3$ is the existence of bifurcation points. Kinematic bifurcation is a situation where the number of mobilities of a mechanism increases at a particular configuration.[1] Here the corresponding configurations are at points (π, π), $(\pi, -\pi)$, $(-\pi, -\pi)$ and $(\pi, -\pi)$. This can be seen in Figure 6.2 as the points where the input–output curves cross. In these configurations, the linkage folds to a bundle where all of the links are

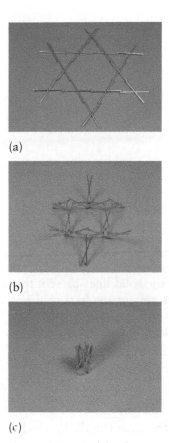

(a)

(b)

(c)

Figure 6.12 Model of a motion assembly of threefold-symmetric Bricard linkages.

collinear to one another. Although bifurcation exists, it does not cause any problem in motion because the links would have to penetrate each other in order to reach the bifurcated position, which is physically impossible.

6.2 Alternative forms of threefold-symmetric Bricard linkage

Although the original threefold-symmetric Bricard linkage can be folded up to a compact bundle, its realisation requires the use of the hinge design introduced in Section 5.7. It has been found however that alternative forms of this linkage also exist similar to the Bennett linkage, which allows much simpler connections among the links.

The method to indentify the alternative form of threefold-symmetric Bricard linkage is the same as that for the Bennett linkage. The axes of

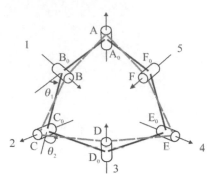

Figure 6.13 The alternative form of threefold-symmetric Bricard linkage.

revolute joints are extended and joints are connected with bars not perpendicular to the axes, see Figure 6.13, where the solid lines present the links of the original linkage and the dashed lines present the bars in the alternative form.

For simplicity, assume that symmetry is kept in the alternative form. We have

$$A_0A = C_0C = E_0E = c,$$
$$B_0B = D_0D = F_0F = d. \tag{6.6}$$

So all the bars of the alternative form have the same length, L, which is

$$AB = BC = CD = DE = EF = FA = L = \sqrt{l^2 + c^2 + d^2 - 2cd\cos\alpha}. \tag{6.7}$$

For each given set of c and d, an alternative form for the threefold-symmetric Bricard linkage can be obtained. The most compact folding can be achieved if simultaneously the points A, C and E meet at a point while the points B, D and F also meet at another point. This means that physically the linkage becomes a bundle whose length is L. Denote θ_1 and θ_2 in this fully folded configuration as θ_{1f} and θ_{2f}, respectively. On the other hand, when the linkage is fully expanded, points A, B, C, D, E and F are on the same plane, i.e. the linkage is completely flattened to form an equilateral hexagon. Take θ_1 and θ_2 in this configuration as θ_{1e} and θ_{2e}.

The relationship among the geometric parameters of the threefold-symmetric Bricard linkage and the design parameters of the linkage in the alternative forms can be found geometrically. Several alternative forms have been studied. Here only two of the most interesting ones, proposed by Pellegrino (2002), are discussed.

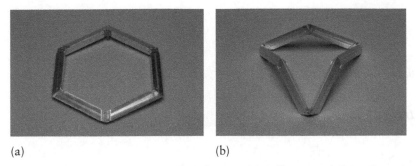

(a)   (b)

Figure 6.14 Motion sequence of Linkage I. (a) Fully expanded configuration and (b) blockage occurs during folding.

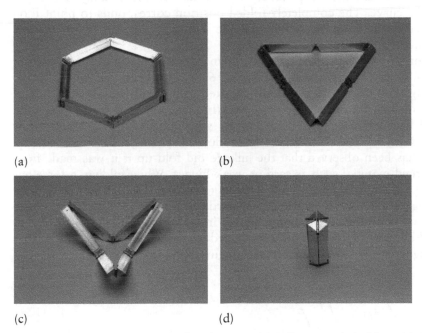

(a)   (b)

(c)   (d)

Figure 6.15 Motion sequence of Linkage II. (a) Fully expanded, (b) to (c) intermediate and (d) fully folded configurations.

A reproduction of the first linkage is shown in Figure 6.14. For convenience, let us call it Linkage I. The other, Linkage II, is shown in Figure 6.15. The geometric parameters of the original threefold-symmetric Bricard linkage and its alternative form of both linkages are as follows (Chen *et al.*, 2005).

$$\alpha = \pi - \arctan 2,$$

$$\theta_{1f} = -\frac{2}{3}\pi, \quad \theta_{2f} = -\arctan\frac{\sqrt{15}}{7},$$

$$\theta_{1e} = \frac{2}{3}\pi, \quad \theta_{2e} = 0, \tag{6.8}$$

$$L = \frac{2}{3}\sqrt{3}\cdot l,$$

$$c = \frac{\sqrt{3}}{6}\cdot l, \quad d = \frac{\sqrt{15}}{6}\cdot l.$$

The input–output curve of both linkages is plotted in Figure 6.16. When the models are fully expanded, they are flat, corresponding to point D on the curve. The completely folded position corresponds to point F on the curve. It is at this point that the similarity of these two linkages ends.

For Linkage I, the folding process can be traced along the input–output curve from D to F via B, see Figure 6.16. However, the movement of linkage is found to be physically blocked at point B because the ends of bars of each hit each other, see Figure 6.14(b). This model is made from solid steel bars connected by brass hinges, which are fairly rigid and allow almost no deformation. Hence, the motion terminates at B.

It has been observed that the linkage did fold up if it was made from less rigid material such as card or weak hinges. While folding, a force has to be applied to linkage to enable joints to deform slightly. A model shown in Figure 6.17, which is made from card, demonstrates this process. After moving from the configuration shown in Figure 6.17(a) to that in Figure 6.17(b) during folding, a force is applied to make the model folded to that in Figure 6.17(c) and then to that in Figure 6.17(d).

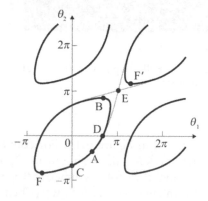

Figure 6.16 θ_2 versus θ_1 curve of the 6R linkage with twist $\alpha = \pi - \arctan 2$.

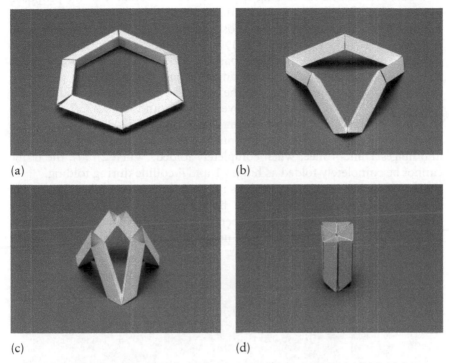

(a)   (b)

(c)   (d)

Figure 6.17 Card model of Linkage I. (a) At D, (b) B, (c) E and (d) F′ of the compatibility path.

A close examination of the input–output curve reveals that, in the card model, the folding process corresponds to a movement from D to F′ of the input–output curve, instead of F, due to the fact that curve is periodic and $\theta_{1F'} = 2\pi + \theta_{1F}$, $\theta_{2F'} = 2\pi + \theta_{2F}$. The reason for this is that $\alpha = \pi - \arctan 2$, i.e. 116.57°, is sufficient close to $2\pi/3$, i.e. 120°. For the $6R$ linkage with twist $2\pi/3$, the input–output curve crosses at the point $\theta_1 = \theta_2 = \pi$. Hence, when the force is applied to the linkage with $\alpha = \pi - \arctan 2$, an imperfection is introduced to the twist of the linkage which alters to $2\pi/3$. The folding process reaches bifurcation point E. When the force is released, the twist of linkage changes back to $\alpha = \pi - \arctan 2$. Accordingly, (θ_1, θ_2) reaches point F′.

Linkage II behaves differently. The folding process takes route from D to F via A and C. There is no blockage during deployment and the structure can be folded up completely, as shown in Figure 6.15.

6.3 Line and plane symmetric Bricard linkage and its alternative forms

Another Bricard linkage found to be useful in creation of motion structures is that with both line and plane symmetry.

Figure 6.18 shows the folding sequence of a *6R* foldable frame proposed by Pellegrino, Green, Guest and Watt (2000), whose geometry is shown in Figure 6.19. Six bars with square cross-section, 1-2, 2-3, 3-4, 4-5, 5-6 and 6-1, are connected by six hinges at 1, 2, 3, 4, 5 and 6, to form a rectangular frame with a symmetric plane passing through hinges 1 and 4. Note that the bars are laid tilted by an angle μ, and the hinges are placed on either the inner or outer surface of the bars. The frame can be folded into a bundle. Links 12, 34, 45 and 61 have length l_1, and links 23 and 56 have length l_2. When $2l_1 < l_2$, the frame can folded and hinges 1 and 4 are a distance apart in the folded configuration. When $2l_1 = l_2$, the frame is a square and hinges 1 and 4 meet when completely folded. When $2l_1 > l_2$, the frame cannot be completely folded as hinges 1 and 4 collide during folding.

This *6R* frame is a linkage with usually one internal mobility. It is not in the original form of a *6R* linkage because in mechanism theory, links always span the shortest distance between two adjacent revolute hinges, whereas here the physical links do not. Thus we name the frame as the

Figure 6.18 Folding sequence of the *6R* foldable frame.

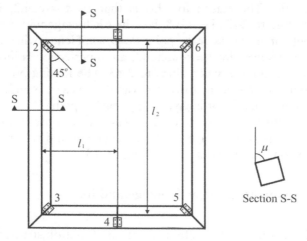

Figure 6.19 The geometry of the *6R* foldable frame.

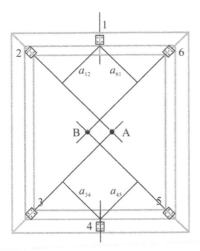

Figure 6.20 The 6R frame and its original linkage.

alternative form of the original linkage. By extending the hinge axes of the six-bar frame, we are able to identify the corresponding links of the original linkage, Figure 6.20.

Because of symmetry of the 6R frame, we have

$$a_{12} = a_{34} = a_{45} = a_{61}. \tag{6.9}$$

Since axes of joints 2 and 3 meet at point A and axes of joints 5 and 6 meet at point B,

$$a_{23} = a_{56} = 0. \tag{6.10}$$

Following the convention given by Beggs (1966), we can define axes x_i and z_i in the original 6R linkage shown in Figure 6.21. Let x_3 at point A be perpendicular to both z_2 and z_3 and point out of the paper, and x_6 at point B be perpendicular to axes z_5 and z_6 and also point out of the paper, as shown in Figure 6.21. So the twists of the linkage are

$$\alpha_{12} = \alpha_{45} = \frac{3\pi}{2}, \quad \alpha_{34} = \alpha_{61} = \frac{\pi}{2}, \tag{6.11a}$$

$$\alpha_{23} = \alpha_{56}, \tag{6.11b}$$

and offsets

$$R_1 = R_4 = 0 \text{ and } -R_2 = R_3 = -R_5 = R_6. \tag{6.12}$$

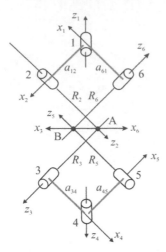

Figure 6.21 The original Bricard linkage. Note that axes x_3 and x_6 are not drawn.

Comparing Eqs (6.10)–(6.12) with Eq. (2.30), it can be concluded that this original linkage is a line-symmetric Bricard linkage with symmetric line passing through the centre and perpendicular to the paper for the configuration in Figure 6.21.

Note that, if either x_3 or x_6 is reversed, which is possible because they correspond to zero length links, the twists α_{23} or α_{56} have to change accordingly. For instance, if x_6 reverses its direction, Eq. (6.11b) has to be replaced by

$$\alpha_{56} = 2\pi - \alpha_{23},$$

which indicates that this original linkage is actually a plane-symmetric Bricard linkage by comparison with Eq. (2.31).

It is therefore concluded that the corresponding original linkage of the 6R frame is a special Bricard linkage with *two-fold symmetry*. Next, we try to find the relationship between the geometric parameters of the original linkage and those of the alternative form.

Because of symmetry, only links 12 and 23 are studied, as shown in Figure 6.22 with both the original linkage and the alternative form. Let $ED = l_1$ and $EG = l_2$. We have

$$\angle DEM = \angle KEM = \arcsin \frac{1}{\sqrt{1 + \cos^2 \mu}}.$$

DM is the link 12 in original linkage, MA is the offset of joint 2 and $\angle FNH$ is the twist of link 23. So we have

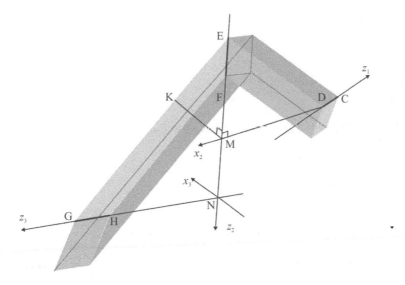

Figure 6.22 The detailed links of a six-bar frame and its original linkage.

$$a_{12} = \frac{l_1}{\sqrt{1+\cos^2\mu}},$$

$$R_2 = \frac{l_2}{2}\frac{\sqrt{1+\cos^2\mu}}{\cos\mu} - l_1\frac{\cos\mu}{\sqrt{1+\cos^2\mu}}, \qquad (6.13)$$

$$a_{23} = \arccos\frac{1-\cos^2\mu}{1+\cos^2\mu}.$$

For the 6R frame, the revolute variables of the original linkage at the fully expanded and fully folded configurations are

$$\theta_{1e} = 2\arcsin\frac{1}{\sqrt{1+\cos^2\mu}} - \pi, \ \theta_{2e} = \frac{\pi}{2} + \arccos(\cos^2\mu),$$

$$\theta_{1f} = -2\arcsin\frac{1}{\sqrt{1+\cos^2\mu}} \ \text{and} \ \theta_{2f} = \frac{3\pi}{2}. \qquad (6.14)$$

Applying the matrix method, the closure equations of this Bricard linkage are obtained as

$$-\tan\frac{\theta_1}{2}\tan\frac{a_{23}}{2} = \sin\theta_2, \qquad (6.15a)$$

$$\theta_3 = \theta_2 + \pi, \qquad (6.15b)$$

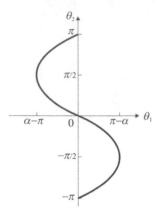

Figure 6.23 The input–output curve for the Bricard linkage.

$$\theta_4 = \theta_1, \ \theta_5 = \theta_2, \ \theta_6 = \theta_3. \tag{6.15c}$$

Figure 6.23 shows the relationship of θ_2 vs θ_1.

However, further analysis has uncovered that the linkage shown in Figure 6.18 always has a bifurcation point as demonstrated in Figure 6.24 (Chen and You, 2009; Chen and Chai, 2011). For the $6R$ frame with $2l_1 \le l_2$ and $0 \le \mu \le \pi/2$,

$$\frac{R_2}{a_{12}} \ge \tan \frac{a_{23}}{2}$$

always holds because of the geometric relationship given by Eq. (6.13). By analysing motion path, it is found that there is at least one bifurcation between fully expanded and folded configurations.

(a)   (b)   (c)

Figure 6.24 The bifurcation of a model frame. (a) Bifurcation point, (b) the configurations of the spherical 4R linkage in which the bottom left and top right corner hinges are frozen and (c) configuration similar to (b) but with other two hinges frozen.

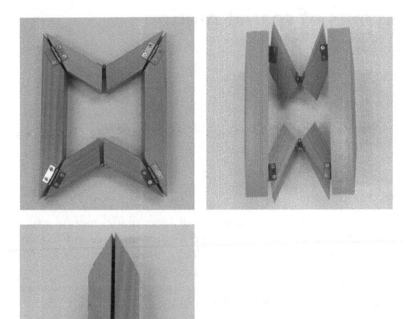

Figure 6.25 The frame without bifurcation.

The bifurcation can be avoided should α_{23} be greater than $\pi/2$ and

$$\cot\frac{\alpha_{23}}{2} < \frac{R_2}{a_{12}} < \tan\frac{\alpha_{23}}{2}.$$

This only becomes possible when the horizontal bars are replaced by inclined bars. Figure 6.25 shows such an example in which

$$\frac{R_2}{a_{12}} = 1$$

and $\alpha_{12} = \alpha_{45} = 3\pi/2$, $\alpha_{23} = \alpha_{56} = 2\pi/3$ and $\alpha_{34} = \alpha_{61} = \pi/2$. As expected no bifurcation is detected in the model. The design retains two-fold symmetry. Details of the bifurcation analysis and the design involving inclined bars can be found in Chen and You (2009).

7 Layouts of spatial motion structures

DOI: 10.1201/9781482266610-7

7.1 Tilings and patterns

One of the most important aspect in design of large spatial motion structures is the identification of a suitable layout. As demonstrated in Chapters 5 and 6, in a chosen layout, the building blocks, often based on a known mechanism, are repeatedly used leading to a generic solution for a type of motion structure. The number of building blocks can be altered depending on the practical size requirement but the mobility of each block is always retained.

Since most of the building blocks, though three dimensional, can be represented by two dimensional polygons schematically, a convenient method for the design of layouts is to utilise a mathematical tool known as tiling, also frequently referred to as tessellation.

A *plane tiling* is a countable family of closed sets which cover the plane without gaps or overlaps. The closed sets are called *tiles* of the tiling. The layout of tiles, termed as a *pattern* in tiling, is a design which repeats some motif in a more or less systematic manner. The art of designing tilings and patterns is clearly extremely old and well developed (Beverley, 1999; Evans, 1931; Rossi, 1970). By contrast, the science of tilings and patterns, which means the study of their mathematical properties, is comparatively recent and many parts of the subject have yet to be explored in depth. The most methodological study of tilings and patterns can be found in Grünbaum and Shephard (1986). Only a brief introduction is provided here.

In mathematics tilings by regular polygons are usually represented by the number of sides of the polygons around any cross point in the clockwise or anti-clockwise order. For instance, (3^6) is a tiling in which each of the points is surrounded by six triangles, '3' is the number of the sides of a triangle and superscript '6' is the number of triangles. Similarly, $(3^3.4^2)$ means three triangles and two squares around a cross-point. And $(3^6;3^2.6^2)$ represents a two-uniform tiling in which there are two types of points, one type is surrounded by six triangles whereas the other type is surrounded by two triangles and two hexagons. The tilings accommodating regular polygons can be classified into four types: regular and uniform tilings, *k*-uniform tilings, equitransitive and edge-transitive tilings, and tilings that are not edge-to-edge (Grünbaum and Shephard, 1986).

The only edge-to-edge monohedral tilings by regular polygons are the three regular tilings shown in Figure 7.1. The basic tiles are identical equilateral triangles, squares and regular hexagons, respectively. There exist precisely eleven distinct edge-to-edge uniform tilings by more than one type of regular polygons such that all vertices are of the same type. They are (3^6), $(3^4.6)$, $(3^3.4^2)$, $(3^2.4.3.4)$, $(3.4.6.4)$, $(3.6.3.6)$, (3.12^2), (4^4), $(4.6.12)$, (4.8^2) and (6^3). An edge-to-edge tiling by regular polygons is called *k*-uniform if its vertices form precisely *k* transitivity classes with respect to the group of symmetries of the tilings. Denote $K(k)$ as the number of distinct *k*-uniform tilings. $K(1)=11$, $K(2)=20$, $K(3)=39$, $K(4)=33$, $K(5)=15$, $K(6)=10$, $K(7)=7$ and $K(k)=0$ when $k \geq 8$. So the total number of distinct *k*-uniform tilings is 135. These tilings can be modified into many more tilings and patterns with methods such as transformation of symmetry, transitivity and regularity, tilings that are not edge-to-edge and patterns with overlap motifs.

The regular and uniform tilings, though simple, can be extended by allowing for each tile itself containing a pattern that differs from the polygonal

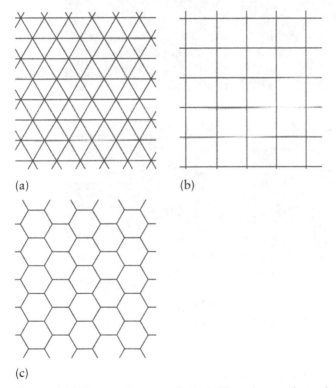

(a)   (b)

(c)

Figure 7.1 Edge-to-edge monohedral tilings by regular polygons. (a) (3^6), (b) (4^4) and (c) (6^3) tilings.

shape of the tile. If a building block, consisting of a single mobile element or a mobile assembly of a number of interconnected elements, has a planar schematic representation that can be inscribed by a triangular, square or hexagonal tile, the corresponding tiling instantly gives a suitable layout provided that the patterns within a tile can be seamlessly connected to those in neighbouring tiles. Figure 7.2 shows a number of possibilities. The remaining task is to ensure that the kinematic requirements can be met in a particular tiling.

In the following sections, the above approach is to be applied to motion structures based on the Bennett, Myard and Bricard linkages.

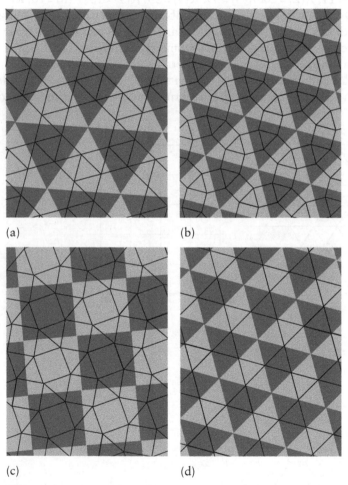

(a)   (b)

(c)   (d)

Figure 7.2 (a)–(g) Repeated patterns surrounded by triangular and square tiles in regular and uniform tilings. Some of them, e.g. (a)–(e) are in fact also edge-to-edge uniform tilings by more than one type of regular polygons

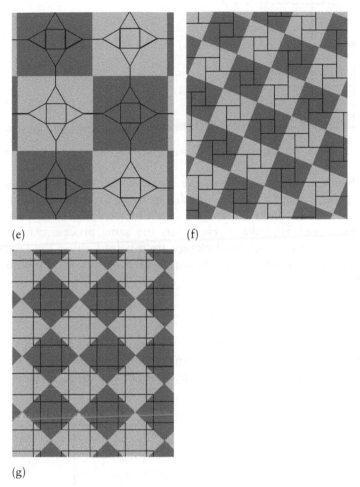

(e)   (f)

(g)

Figure 7.2 continued

7.2 Layouts for the Bennett linkage

7.2.1 *Case I*

A single Bennett linkage consisting of four links can be represented schematically by a square as in Chapter 5. When it is used as a building block inscribed by a square of identical orientation, tiling (4^4) in Figure 7.1(b) presents the simplest way to form an assembly in which other Bennett linkages are connected to each of its sides completing a layout shown in Figure 7.3(a).

The kinematics of the assembly can be examined by first considering two adjacent Bennett linkages ABCD and DCEF, Figure 7.3. Without other

Figure 7.3 Several Bennett linkages connected side by side.

Bennett linkages connected to them the two together would have more than one degree of mobility because each can move independently. Goldberg (1943) found that links BC and CF could be welded together, resulting in the Goldberg 5R linkage. However, the same process cannot be repeated if other adjacent Bennett linkages are involved unless the Bennett linkages degenerate to planar 4R linkages. Hence, this assembly is in general not mobile.

7.2.2 Case II

Should the layout given in Figure 7.2(f) be selected, the building block inscribed by the square tile is a Bennett linkage with extended links, Figure 7.4(a), and if the revolute joints are kept at A, B, C and D, an assembly

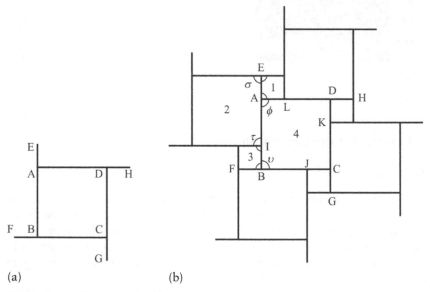

(a)   (b)

Figure 7.4 (a) A Case II building block based on the Bennett linkage and (b) assembly of the building blocks.

can then be formed, Figure 7.4(b). Next, let us examine kinematically the mobility of such an assembly.

Let

$$a_{AB} = a_{CD} = a, \; a_{BC} = a_{DA} = b,$$

$$\alpha_{AB} = \alpha_{CD} = \alpha, \; \alpha_{BC} = \alpha_{DA} = \beta,$$

$$a_{AE} = a_{CG} = a_{BI} = a_{DK} = k_s a, \; a_{AL} = a_{CJ} = a_{BF} = a_{DH} = k_s b,$$

$$\alpha_{AE} = \alpha_{CG} = \alpha_{BI} = \alpha_{DK} = \alpha, \; \alpha_{AL} = \alpha_{CJ} = \alpha_{BF} = \alpha_{DH} = \beta, \tag{7.1}$$

where k_s is a constant with $0 < k_s < 1$. All of squares in Figure 7.4(b) have zero offsets. Hence, they satisfy the geometric conditions for the Bennett linkage, i.e. Eqs (2.23)–(2.25).

Considering Bennett linkages 1, 2, 3 and 4, there are

$$\tan \frac{\sigma}{2} \tan \frac{\phi}{2} = \frac{\sin \frac{1}{2}(\beta + \alpha)}{\sin \frac{1}{2}(\beta - \alpha)}, \tag{7.2a}$$

$$\tan \frac{\pi - \sigma}{2} \tan \frac{\pi - \tau}{2} = \frac{\sin \frac{1}{2}(\beta + \alpha)}{\sin \frac{1}{2}(\beta - \alpha)}, \tag{7.2b}$$

$$\tan \frac{\tau}{2} \tan \frac{\upsilon}{2} = \frac{\sin \frac{1}{2}(\beta + \alpha)}{\sin \frac{1}{2}(\beta - \alpha)}, \tag{7.2c}$$

and

$$\tan \frac{\pi - \phi}{2} \tan \frac{\pi - \upsilon}{2} = \frac{\sin \frac{1}{2}(\beta + \alpha)}{\sin \frac{1}{2}(\beta - \alpha)}. \tag{7.2d}$$

Combining Eqs (7.2a) and (7.2c), as well as (7.2b) and (7.2d), respectively, gives

$$\tan \frac{\phi}{2} \tan \frac{\sigma}{2} \tan \frac{\tau}{2} \tan \frac{\upsilon}{2} = \left(\frac{\sin \frac{1}{2}(\beta + \alpha)}{\sin \frac{1}{2}(\beta - \alpha)} \right)^2, \tag{7.3a}$$

$$\tan\frac{\phi}{2}\tan\frac{\sigma}{2}\tan\frac{\tau}{2}\tan\frac{\upsilon}{2}=\left(\frac{\sin\frac{1}{2}(\beta-\alpha)}{\sin\frac{1}{2}(\beta+\alpha)}\right)^{2}. \tag{7.3b}$$

From Eqs (7.3a) and (7.3b), we obtain that

$$\frac{\sin\frac{1}{2}(\beta-\alpha)}{\sin\frac{1}{2}(\beta+\alpha)}=\pm1. \tag{7.3}$$

The solutions to Eq. (7.3) are, for any $0\leq\alpha\leq2\pi$ and $0\leq\beta\leq2\pi$,

$$\beta-\alpha=\beta+\alpha,$$

$$\beta-\alpha=-(\beta+\alpha),$$

$$\beta+\alpha=(\beta-\alpha)+2\pi,$$

$$\beta+\alpha=-(\beta-\alpha)+2\pi,$$

leading to

$$\alpha=0\text{, or }\alpha=\pi\text{, and }\beta=0\text{, or }\beta=\pi. \tag{7.4}$$

It can be concluded from Eq. (7.4) that only when α and β are 0 or π can the assembly based on the tiling shown in Figure 7.4(b) remain mobile. The Bennett linkage now degenerates to a planar 4R linkage for Bennett linkages with twists other than 0 or π the mobility vanishes.

7.2.3 Case III

Now modify the building block further to that shown in Figure 7.5(a), in which all of the links are extended at both ends. Using the pattern of Figure 7.2(g) produces the assembly shown in Figure 7.5(b), which is identical to the layout given in Section 5.2.1. Hence, the assembly is mobile if the conditions given in Section 5.2.2 are met.

7.2.4 Case IV

This is an extension of Case III. If only two corners of the Bennett linkage are extended, Figure 7.6(a), the assembly shown in Figure 7.6(b) can be obtained. Let the Bennett linkage ABCD have lengths a, b and twists α, β

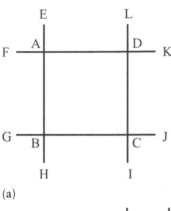

(a)

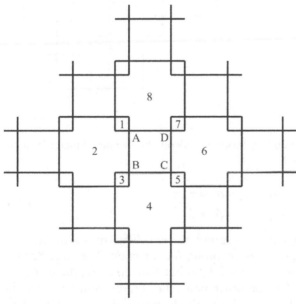

(b)

Figure 7.5 (a) A Case III building block based on the Bennett linkage and (b) assembly of the building blocks.

and the interconnected Bennett linkages around it, which are marked with numbers 1 to 8, have lengths a_i, b_i and twists α_i, β_i ($i = 1, 2, \ldots, 8$). It can be found that the mobility conditions for this assembly are as follows (Chen and You, 2008b).

$$\alpha_1 = \alpha_5 = -\alpha, \quad \alpha_2 = -\alpha_3 = \alpha_4, \quad \alpha_6 = -\alpha_7 = \alpha_8,$$
$$\beta_1 = \beta_5 = -\beta, \quad \beta_2 = -\beta_3 = \beta_4, \quad \beta_6 = -\beta_7 = \beta_8,$$

(7.5a)

or

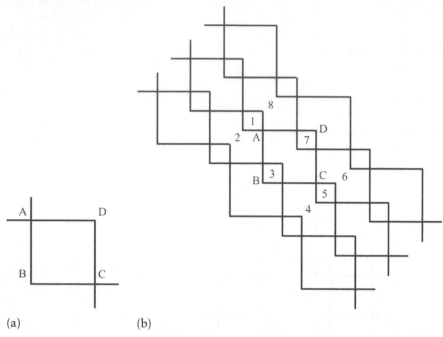

Figure 7.6 (a) A Case IV building block based on the Bennett linkage and (b) assembly of blocks.

$$\alpha_3 = \alpha_7 = \alpha, \quad \alpha_1 = \alpha_2 = \alpha_8, \quad \alpha_4 = \alpha_5 = \alpha_6,$$
$$\beta_3 = \beta_7 = \beta, \quad \beta_1 = \beta_2 = \beta_8, \quad \beta_4 = \beta_5 = \beta_6. \tag{7.5b}$$

Eqs (7.5a) and (7.5b) in fact correspond to two different assemblies. Eq. (7.5a) indicates that the guidelines are along AC direction. The large Bennett linkages do not overlap along the guidelines but they do along the diagonals parallel to BD direction. On the other hand, Eq. (7.5b) points to that the guidelines are along BD direction. Again the Bennett linkages overlap along the guidelines, but not so along the diagonals parallel to AC direction.

Nevertheless the lengths of each Bennett linkage in both assemblies must satisfy

$$\frac{\sin \alpha}{\sin \beta} = \frac{\sin \alpha_i}{\sin \beta_i} = \frac{a}{b} = \frac{a_i}{b_i} \quad (i = 1, 2, \ldots, 8), \tag{7.6}$$

because of Eq. (2.24).

The expansion sequence of models of the two mobile assemblies in this case are given Figures 7.7(a) and (b), respectively.

Case IV assemblies can also be regarded as a special case of Case III if we consider that the length of a link can have negative value. Readers may refer to Chen and You (2008b) for detailed discussion.

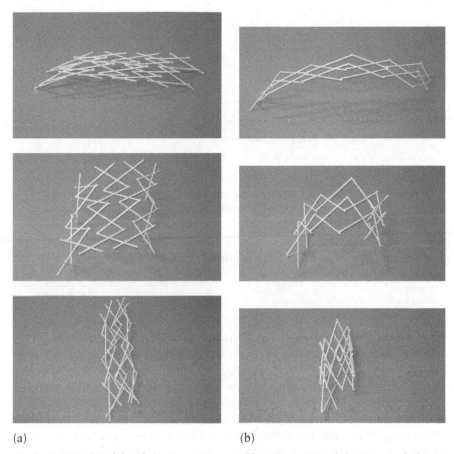

Figure 7.7 (a) Models of the Case IV assembly. The twists of the Bennett linkages are governed by (a) Eq. (7.5a) and (b) Eq. (7.5b), respectively.

7.3 Assemblies of Myard linkages

7.3.1 Building blocks

The Myard linkage (Myard, 1931; Baker, 1979) consists of five bars, one of which has zero length. An umbrella-shaped motion structure, Figure 7.8, was reported consisting of a series of circumferentially connected Myard linkages (Briand and You, 2007). When the assembly consists of n identical Myard linkages, there are n joints at the centre point. Because the frame forms a flat polygon when fully expanded with each Myard linkage forming a triangle, the twist of each Myard linkage at centre must be $\alpha_{12} = \pi/n$. The assemblies with $n = 3$, 4 and 6 can be represented schematically by a triangle, square and hexagon, respectively, which can be used as

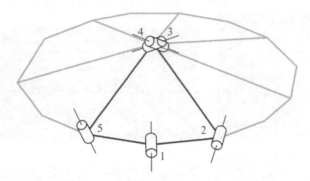

Figure 7.8 A schematic diagram for an umbrella-like deployable frame made from seven Myard linkages. Only one of the Myard linkages is highlighted.

the building blocks to form large motion structures following (3^6), (4^4) and (6^3) tilings (Liu and Chen, 2009).

7.3.2 Case I

The building block composed of three Myard linkages with $\alpha_{12} = \pi/3$ expands to a shape that can be inscribed by a tile whose shape is an equilateral triangle, Figure 7.9(a). Hence, (3^6) tiling can be used to obtain a layout. A number of the same blocks can be assembled and they are connected through the continuous crossbars, Figure 7.9(b). The assembly is mobile because all of the Myard linkages are identical and they move synchronously. The centres of the building blocks move upwards or downwards when the assembly is placed on a horizontal surface, that is, in

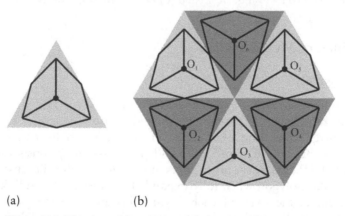

(a)   (b)

Figure 7.9 The assembly of Myard linkages with $\alpha_{12} = \pi/3$. (a) An assembly of three such linkages as a unit and (b) assembly with six units.

Figure 7.9(b), O_1, O_3 and O_5 move upwards together, whereas O_2, O_4 and O_6 move downwards together.

7.3.3 Case II

Should the building block have four Myard linkages with $\alpha_{12} = \pi/4$, the tile is a square and (4^4) tiling becomes a suitable layout, Figure 7.10. Again the building blocks are connected by extended links forming cross bars. The building blocks move synchronously with the centres in the squares with the light shade going upwards whereas those in dark shades downwards if the assembly is placed on a horizontal surface.

7.3.4 Case III

Six Myard linkages with $\alpha_{12} = \pi/6$ can be put together to form an umbrella frame shown in Figure 7.11(a). However, it turned out that this assembly cannot be used directly as a building block with (6^3) tiling as we have done with the previous two cases because the kinematic conditions cannot be satisfied. Certain modification has to be made. Figure 7.11(b) is the umbrella assembly with six additional Myard linkages identical to those forming the umbrella which are joined to the umbrella by pairs of cross-bars. The assembly still has mobility one. When centre C moves upwards during motion, C_i ($i = 1, 2, \ldots, 6$) move downwards. Using this as the building block that is inscribed by a hexagonal tile and following tiling (6^3), three blocks can be assembled together, Figure 7.11(c). The resulting assembly is mobile but has more than one degree of mobility because three Myard linkages meeting at point A can move independently. In order to

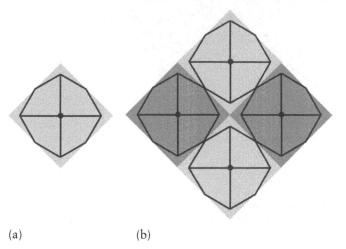

(a)   (b)

Figure 7.10 The assembly of Myard linkages with $\alpha_{12} = \pi/4$. (a) An assembly of four such linkages as a unit and (b) assembly with four units.

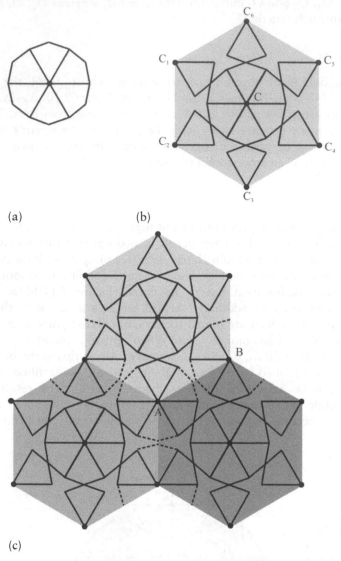

(a)   (b)

(c)

Figure 7.11 Assembly of Myard linkage with $\alpha_{12} = \pi/6$. (a) An assembly of six such linkages and (b) a unit and (c) assembly with three units.

synchronise their motions three pairs of bars, shown in dash line in Figure 7.11(c), are added so that there are a total of six Myard linkages around point A to form an umbrella frame. Now the entire assembly has mobility one. This construction process can be repeated to form a large motion structure.

7.4 Assemblies of threefold-symmetric Bricard linkages

7.4.1 *Case I*

The threefold-symmetric Bricard linkage has six links of equal length. It is natural to use (6^3) tiling for the formation of motion structure.

Consider a portion of the assembly shown in Figure 7.12(a). Loops 1, 2 and 3 are all made of identical threefold-symmetric Bricard linkages with twists of α and $2\pi - \alpha$ alternating from one link to the other. Let

$$\alpha_{AC} = \alpha. \tag{7.7}$$

Then for loop 2,

$$\alpha_{BA} = 2\pi - \alpha. \tag{7.8}$$

Considering loop 3 gives

$$\alpha_{AD} = 2\pi - \alpha. \tag{7.9}$$

Now a problem arises, as for loop 1,

$$\alpha_{BA} = \alpha_{AD} = 2\pi - \alpha, \tag{7.10}$$

meaning that loop 1 violates the condition of the threefold-symmetric Bricard linkage given in Eq. (6.1). Thus, the assembly consisting of more than one threefold-symmetric Bricard linkage will become immobile.

This layout can be revived by using a 6R loop which has the same twist for all of the six links, either α or $2\pi - \alpha$ for loop 1 in the centre while keep the rest the same. The 6R loop at 1 is in fact a very special line and plane symmetric Bricard linkage that has mobility two, resulting in a discontinuous

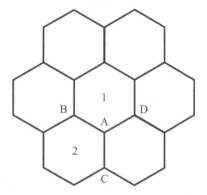

Figure 7.12 An assembly of threefold-symmetric Bricard linkage.

motion. The kinematic motion requirement can now be met. However, the entire assembly no longer exhibits the synchronised motion common for assemblies of mobility one.

7.4.2 Case II

In Section 6.4 we showed that the threefold-symmetric Bricard linkage expands to a triangular shape shown in Figure 7.13(a). Thus, it may be possible to use (3^6) tiling to construct an assembly.

A typical connection of threefold-symmetric Bricard linkages in the (3^6) tiling is as shown in Figure 7.13(b) in which the central linkage shares two common links with each of the adjacent linkages. The projections of its probable expansion sequence are shown in Figures 7.13(c) and (d). It becomes obvious that the central linkage behaves completely differently from those around it. Hence, such an arrangement cannot be repeatedly used to construct a mobile assembly.

7.4.3 Case III

The threefold-symmetric Bricard linkage can be modified by extending the links into cross bars to form the building block shown in Figure 7.14 so that it can be tessellated by (3^6) tiling to make a mobile assembly, which is similar to that presented in Section 6.1.

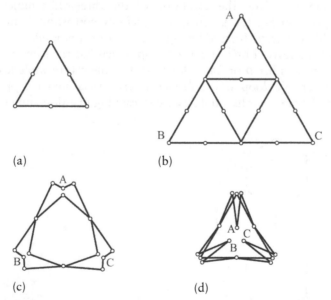

(a)   (b)

(c)   (d)

Figure 7.13 (a) A threefold-symmetric Bricard linkage with twists of $\pi/3$ or $2\pi/3$, (b) possible connections, (c) and (d) projection of probable deployment sequence.

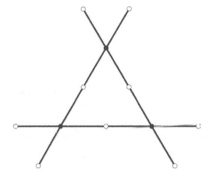

Figure 7.14 A building block based on the threefold-symmetric Bricard linkage with twists of $\pi/3$ or $2\pi/3$.

7.5 Conclusion and discussion

A tessellation method to build motion structures using repeated building blocks and tiling technique is presented in this chapter. It involves three steps: selection of suitable tilings, construction of building blocks using common or overconstrained mechanisms and validation of geometrical compatibility. The first two steps have to be taken jointly and it may possibly involve many design iterations.

This method has been applied to the assemblies consisting of Bennett linkages, Myard linkages and threefold-symmetric Bricard linkages, respectively. All three regular tilings (3^6), (4^4) and (6^3), have been adopted and a number of building blocks of different types have been discussed. Some of the motion structures that have been described in previous chapters have been reconstructed by this tessellation method.

It should be pointed out that there are many ways of constructing building blocks, including those made from a combination of more than one type of three dimensional overconstrained linkages. Great care has to be taken in design of a building block as its motion could have profound influence on the overall performance of the assembly. For instance, if compact folding is the objective, the building blocks themselves have to exhibit this property. Furthermore, once a design is obtained attention has to be paid to the size and shape of the links and details of the joints for a desirable folding configuration may not be realised due to intersection of links.

Only regular and uniform tilings and edge-to-edge tilings are utilised here. Nevertheless this approach seems to yield useful results judging by the successful examples illustrated here. Readers are encouraged to explore other tilings and patterns that can be found from references such as Grünbaum and Shephard (1986).

Note that some of the motion structures with layouts based on tilings exhibit three dimensional motion and have curved profiles despite that tilings are essentially two dimensional. This is due to the fact that the building blocks are based on three dimensional mechanisms. A truly three dimensional layout may be obtained using the uniform polyhedra. In the past, Kovács *et al.* (2004) proposed a class of expandable polyhedral structures transforming between dodecahedron and truncated icosahedrons. More recently, eight threefold-symmetric Bricard linkages are connected with the layout of an octahedron, resulting in a spatial motion structure whose motion follows the transformation between octahedron and cuboctahedron, Figure 7.15 (Cahyono and Chen, 2009). Such motion structures are likely to expand radially forming deployable cubes or spheres. It is much harder to retain the mobility of interlinked building blocks in a three dimensional layout because of the strict constraints in maintaining the shape of a polyhedron, which is why the number of successful examples has so far been rather limited.

Figure 7.15 Transformation between octahedron and cub-octahedron with motion structures based on threefold-symmetric Bricard linkages.

Notes

2 Fundamental concepts, methods and classification

1 Flexistar 6 is produced by the Orb Factory Ltd, Halifax, NS, Canada.

3 Planar double chain linkages

1 In the Hoberman sphere, AE = BE. It is not always necessary
2 Some diagrams shown in this chapter contain μ and ν which are anti-clockwise. They are therefore negative in value.
3 Readers may follow a different route in writing down the vector form of the closure condition. However, it always ends up with Eq. (3.10) because of the loop parallelogram constraint.
4 Some diagrams shown in this chapter contain θ which are anti-clockwise. It is therefore negative in value.

4 Spatial rings and domes

1 Further concepts could be generated in a similar way.

6 Spatial motion structures based on Bricard linkages

1 The increased mobility can be infinitesimal as well as finite.

References

Altmann, P. G. (1954) Communications to Grodzinski, P. and M'Ewen, E., Link mechanisms in modern kinematics. *Proceedings of the Institution of Mechanical Engineers*, 168(37), 889–896.

Altmann, S. L. (1986) *Rotations, quaternions, and double groups*, Oxford University Press, Oxford.

Baker, J. E. (1975) The Delassus linkages. *Proceedings of Fourth World Congress on the Theory of Machines and Mechanisms*, Newcastle upon Tyne, England, 8–13 September, paper No. 9, 45–49.

Baker, J. E. (1979) The Bennett, Goldberg and Myard linkages – in perspective. *Mechanism and Machine Theory*, 14(4), 239–253.

Baker, J. E. (1980) An analysis of the Bricard linkages. *Mechanism and Machine Theory*, 15(4), 267–286.

Baker, J. E. (2002) Displacement-closure equations of the unspecialised double-Hooke's-joint linkage. *Mechanism and Machine Theory*, 37(10), 1127–1144.

Baker, J. E. and Hu, M. (1986) On spatial networks of overconstrained linkages. *Mechanism and Machine Theory*, 21(5), 427–437.

Baker, J. E. and Yu, H. C. (1983) Re-examination of a Kempe linkage. *Mechanism and Machine Theory*, 18(1), 7–21.

Ball, R. S. (1876) *The theory of screws: A study in the dynamics of a rigid body*, Hodges Foster.

Beggs, J. S. (1966) *Advanced Mechanism*, Macmillan Company, New York.

Bennett, G. T. (1903) A new mechanism. *Engineering*, 76, 777–778.

Bennett, G. T. (1914) The skew isogram mechanism. *Proceeding of London Mathematics Society*, 2nd series, 13, 151–173.

Beverley, D. (1999) *Tiling and mosaics in a weekend*, Merehurst, London.

Briand, S. and You, Z. (2007) New deployable mechanisms, Report 2293/07, Department of Engineering Science, University of Oxford.

Bricard, R. (1897) Mémoire sur la théorie de l'octaedre articulé. *Journal de mathématiques pures et appliquées*, Liouville 3, 113–148.

Bricard, R. (1927) Leçons de cinématique. *Tome II Cinématique Appliquée*, Gauthier-Villars, Paris, 7–12.

Cahyono, E. and Chen, Y. (2009) Smart structure of Kaleidocycle. Proceedings of the URECA@NTU 2008–09.

Calladine, C. R. (1978) Buckminster Fuller's 'tensegrity' structures and Clerk Maxwell's rules for the construction of stiff frames. *International Journal of Solids and Structures*, 14(2), 161–172.

Calladine, C. R. and Pellegrino, S. (1991) First-order infinitesimal mechanisms. *International Journal of Solids and Structures*, 27(4), 505–515.

Chen, Y. and Baker, J. E. (2005) On using a Bennett linkage as a connector between other Bennett loops. *Proceeding of Institution of Mechanical Engineers, Journal of Multi-body Dynamics*, 219(2), 177–185.

Chen, Y. and Chai, W. H. (2011) Bifurcation of a special line and plane symmetric Bricard linkage. *Mechanism and Machine Theory*, 46(4): 515–533.

Chen, Y. and You, Z. (2002) Connectivity of Bennett Linkages, AIAA 2002–1500. *Proceeding of 43rd AIAA/ASME/ASCE/AHS/ASC Structures, Structural Dynamics, and Material Conference and Exhibit*, Denver, CO, 22–25 April.

Chen, Y. and You, Z. (2005) Mobile assemblies based on the Bennett linkage. *Proceedings of the Royal Society A (Mathematical, Physical and Engineering Sciences)*, 461(2056), 1229–1245.

Chen, Y. and You, Z. (2006) Square deployable frame for space application: Part I: theory. *Proceedings of the Institution of Mechanical Engineers, Part G, Journal of Aerospace Engineering*, 220(4), 347–354.

Chen, Y. and You, Z. (2007a) Square deployable frame for space application: Part II: realization. *Proceedings of the Institution of Mechanical Engineers, Part G: Journal of Aerospace Engineering*, 221(1), 37–45.

Chen, Y. and You, Z. (2007b) Spatial 6R linkages based on the combination of two Goldberg 5R linkages. *Mechanism and Machine Theory* 42(11), 1484–1498.

Chen, Y. and You, Z. (2007c) Deployable frames with curved profile, AIAA-2007–2115. *Proceeding of 48th AIAA/ASME/ASCE/AHS/ASC Structures, Structural Dynamics, and Material Conference and Exhibit*, Honolulu, Hawaii, 23–26 April.

Chen, Y. and You, Z. (2008a) An extended Myard linkage and its derived 6R linkage. *ASME Journal of Mechanical Design*, 130(5).

Chen, Y. and You, Z. (2008b) On mobile assemblies of Bennett linkages. *Proceedings of the Royal Society A (Mathematical, Physical and Engineering Sciences)*, 464(2093), 1275–1283.

Chen, Y. and You, Z. (2009) Two-fold symmetrical 6R foldable frames and their bifurcations. *International Journal of Solids and Structures*, 46(25–26), 4504–4514.

Chen, Y., You, Z. and Tarnai, T. (2005) Threefold-symmetric Bricard linkages for deployable structures. *International Journal of Solids and Structures*, 42(8), 2287–2301.

Crawford, R. F., Hedgepeth, J. M. and Preiswerk, P. R. (1973) Spoked wheels to deploy large surfaces in space: Weight estimates for solar arrays, *NASA-CR-2347*.

Denavit, J. and Hartenberg, R. S. (1955) A kinematic notation for lower-pair mechanisms based on matrices. *Journal of Applied Mechanics*, 22(2), 215–221.

Dietmaier, P. (1995) A new 6R space mechanism. *Proceedings 9th World Congress IFToMM*, Milano, 1, 52–56.

Escrig, F., Valcarcel, J. P. and Sanchez, J. (1996) The adventure of covering a swimming-pool with an X-frame structure. in: Escrig, F. and Brebbia, C. eds, *Mobile and Rapidly Assembled Structure II*, Wessex Institute of Technology, UK, and University of Seville, Spain, 113–122.

Evans, J. (1931) *Pattern: A study of ornament in western Europe from 1180 to 1900*, Clarendon Press, Oxford.

Gantes, C. (1991) A design methodology for deployable structures, PhD Thesis, Massachusetts Institute of Technology.

Goldberg, M. (1943) New five-bar and six-bar linkages in three dimensions, *Trans. ASME*, 65, 649–663.

Grünbaum, B. and Shephard, G. C. (1986) *Tilings and patterns*, W. H. Freeman and Company, New York.

Hartenberg, R. S. and Denavit, J. (1964) *Kinematic synthesis of linkages*, McGraw-Hill Book Company, New York.

Hoberman, C. (1990) Reversibly expandable doubly-curved truss structures. US Patent 4,942,700.

Hoberman, C. (1991) Radial expansion retraction truss structure. US Patent 5,024,031.

Hunt, K. H. (1978) *Kinematic geometry of mechanisms*, Oxford University Press, Oxford.

Ishii, K. (2000) *Structural design of retractable roof structures*, WIT Press, Southampton.

Jensen, F. and Pellegrino, S. (2002) Expandable structures formed by hinged plates. In: *Proceedings Fifth International Conference on Space Structures*, Thomas Telford Limited, Guildford, Surrey, UK, 19–21.

Kassabian, P. E., You, Z. and Pellegrino, S. (1999) Retractable roof structures. *Proceedings of the Institute of Civil Engineers: Structures and Buildings*, 134(1), 45–56.

Kempe, A. B. (1878) On conjugate four-piece linkages. *Proceeding of London Mathematics Society*, 9, 133–157.

Kovács, F., Tarnai, T., Fowler, P. W. and Guest, S. D. (2004) A class of expandable polyhedral structures. *International Journal of Solids and Structures*, 41(3–4), 1119–1137.

Kuipers, J. (2002) *Quaternions and rotation sequences: A primer with applications to orbits, aerospace, and virtual reality*, Princeton University Press, Princeton, NJ.

Lee, C.-C. and Dai, J. S. (2003) Configuration analysis of the Schatz linkage. *Journal Proceedings of the Institution of Mechanical Engineers, Part C: Journal of Mechanical Engineering Science*, 217(7), 779–786.

Liu, S. Y. and Chen, Y. (2009) Myard linkage and its mobile assemblies. *Mechanism and Machine Theory*, 44(10), 1950–1963.

Luo, Y., Mao, D. and You, Z. (2007) On a type of radially retractable plate structures. *International Journal of Solids and Structures*, 44(10), 3452–3467.

McCarthy, J. M. (1990) *An introduction to theoretical kinematics*, MIT Press, Cambridge, MA.

Mao, D., Luo, Y. and You, Z. (2009) Planar closed loop double chain linkages. *Mechanisms and Machine Theory*, 44(4), 850–859.

Mavroidis, C. and Roth, B. (1994) Analysis and synthesis of overconstrained mechanism. *Proceeding of the 1994 ASME Design Technical Conference*, Minneapolis, MI, September, 115–133.

Meurant, R. (1993) Circular space trusses. In: Parke, G. A. R. and Howard, C. M. eds, *Space Structures 4*, Thomas Telford, London, 2053–2062.

Myard, F. E. (1931) Contribution à la géométrie des systèmes articulés. *Societe mathématiques de France*, 59, 183–210.

Pellegrino, S. (2002) Personal communication.

Pellegrino, S. and Calladine, C. R. (1986) Matrix analysis of statically and kinematically indeterminate frameworks. *International Journal of Solids and Structures*, 22(4), 409–428.

Phillips, J. (1990) *Freedom of machinery, volume II*, Cambridge University Press, Cambridge.

Reuleaux, F. (1875) *The kinematics of machinery*, Macmillan and Co., London.

Rossi, F. (1970) *Mosaics: A survey of their history and techniques*, Pall Mall Press, London.

Sánchez-Cuenca, L. (1996) Geometric models for expandable structures. In: Escrig, F. and Brebbia, C. eds, *Mobile and Rapidly Assembled Structure II*, Computational Mechanics Publications, 93–102.

Sarrus, P. T. (1853) Note sur la transformation des mouvements rectilignes alternatives, en mouvements circulaires, et reciproquement. *Académie des Sciences*, 36, 1036–1038.

Savage, M. (1972) Four-link mechanisms with cylindric, revolute and prismatic pairs. *Mechanism and Machine Theory*, 7(2), 191–210.

Schattschneider, D. and Walker, W. (1977) *M. C. Escher kaleidocycles*, Tarquin Publications, St Albans, England.

Tarnai, T. (1984) Bifurcation of equilibrium and bifurcation of compatibility. In: Abstracts of Lectures. *16th International Congress of Theoretical and Applied Mechanics*, Lyngby, Denmark. Lecture 652.

Tarnai, T. (2001) Infinitesimal and finite mechanisms. In: Pellegrino, S. ed. *Deployable Structures*, CISM Courses and Lectures No. 412. Springer, Wien, New York, 113–142.

Tian, P. L. and Chen, Y. (2010) Design of a foldable shelter. *Proceedings of the International Symposium of Mechanism and Machine Theory*, AzCIFToMM – Izmir Institute of Technology, Izmir, Turkey, 5–8 October.

Varadarajan, V. S. (1974) *Lie groups, Lie algebras, and their representations*, Prentice-Hall, Upper Saddle River, NJ.

Waldron, K. J. (1968) Hybrid overconstrained linkages. *Journal of Mechanisms*, 3(2), 73–78.

Wohlhart, K. (1987) A new 6R space mechanism. *Proceedings of the 7th World Congress on the Theory of Machines and Mechanisms*, Seville, Spain, 17–22 September, 1, 193–198.

Wohlhart, K. (1991) Merging two general Goldberg 5R linkages to obtain a new 6R space mechanism. *Mechanism and Machine Theory*, 26(2), 659–668.

Wohlhart, K. (2000) Double-chain mechanisms. In: Pellegrino, S. and Guest, S. D. eds, *IUTAM-IASS Symposium on Deployable Structures: Theory and Applications*, Kluwer Academic Publications, Dordrecht, the Netherlands, 457–466.

You, Z. and Pellegrino, S. (1997a) Foldable bar structures. *International Journal of Solids and Structures*, 34(15), 1825–1847.

You, Z. and Pellegrino, S. (1997b) Cable-stiffened pantographic deployable structures. Part 2: Mesh Reflector. *AIAA Journal*, 35(8), 1348–1355.

Zeigler, T. R. (1981) Collapsible self-supporting structures and panels and hub therefor, US Patent 4,290,244.

Index

Milton Keynes UK
Ingram Content Group UK Ltd.
UKHW040052071024
449327UK00019B/515